Shiv Naresh Shivhare, Thipendra P. Singh, Deepa Joshi, Anitesh Mishra

Computer-Aided Intelligent Imaging

Also of Interest

Digital Transformation in Healthcare 5.0
Volume 1: IoT, AI and Digital Twin
Rishabha Malviya, Sonali Sundram, Rajesh Kumar Dhanaraj, Seifedine Kadry (Eds.), 2024
ISBN 978-3-11-132646-7, e-ISBN (PDF) 978-3-11-132785-3,
e-ISBN (EPUB) 978-3-11-132786-0

Digital Transformation in Healthcare 5.0
Volume 2: Metaverse, Nanorobots and Machine Learning
Rishabha Malviya, Sonali Sundram, Rajesh Kumar Dhanaraj, Seifedine Kadry (Eds.), 2024
ISBN 978-3-11-139738-2, e-ISBN (PDF) 978-3-11-139854-9,
e-ISBN (EPUB) 978-3-11-139911-9

Healthcare Big Data Analytics
Computational Optimization and Cohesive Approaches
Volume 10 of the series Intelligent Biomedical Data Analysis
Akash Kumar Bhoi, Ranjit Panigrahi, Victor, Hugo C. de Albuquerque, Rutvij H. Jhaveri (Eds.), 2024
ISBN 978-3-11-075073-7, e-ISBN (PDF) 978-3-11-075094-2,
e-ISBN (EPUB) 978-3-11-075098-0

Artificial Intelligence and Computational Dynamics for Biomedical Research
Volume 8 of the series Intelligent Biomedical Data Analysis
Ankur Saxena, Nicolas Brault (Eds.), 2022
ISBN 978-3-11-076199-3, e-ISBN (PDF) 978-3-11-076204-4,
e-ISBN (EPUB) 978-3-11-076208-2

Artificial Intelligence for Data-Driven Medical Diagnosis
Volume 3 of the series Intelligent Biomedical Data Analysis
Deepak Gupta, Utku Kose, Bao Le Nguyen, Siddhartha Bhattacharyya (Eds.), 2021
ISBN 978-3-11-066781-3, e-ISBN (PDF) 978-3-11-066832-2,
e-ISBN (EPUB) 978-3-11-066838-4

Shiv Naresh Shivhare, Thipendra P. Singh,
Deepa Joshi, Anitesh Mishra

Computer-Aided Intelligent Imaging

Artificial Intelligence in Healthcare Informatics

DE GRUYTER

Authors
Dr. Shiv Naresh Shivhare
Tower-C, Gaur Atulyam, Omicron 1
201310 Greater Noida
Uttar Pradesh
India
shiv827@gmail.com

Dr. Thipendra P. Singh
Millennium Village, Alpha -I, 35
201310 Greater Noida
Uttar Pradesh
India
thipendra@gmail.com

Dr. Deepa Joshi
Campus Hub, Triq Roberto Ranieri Costaguti
L-Imsida MSD 2080
Malta (Europe)
deepajoshi117@gmail.com

Anitesh Mishra
Tower-J, Gaur Atulyam, Omicron 1
201310 Greater Noida
Uttar Pradesh
India
anitesh.mishra@yahoo.co.in

ISBN 978-3-11-138902-8
e-ISBN (PDF) 978-3-11-138905-9
e-ISBN (EPUB) 978-3-11-138934-9

Library of Congress Control Number: 2026930396

Bibliographic information published by the Deutsche Nationalbibliothek
The Deutsche Nationalbibliothek lists this publication in the Deutsche Nationalbibliografie; detailed bibliographic data are available on the Internet at http://dnb.dnb.de.

Cover image: alvarez / E+ / Getty Images
Typesetting: VTeX UAB, Lithuania

www.degruyterbrill.com
Questions about General Product Safety Regulation:
productsafety@degruyterbrill.com

To our families, for their love and patience.

And to the healthcare community and patients, whose resilience inspires this work

Preface

This book brings together recent advances in medical imaging with a focus on artificial intelligence and computer-aided diagnostic systems. Our aim is to provide a coherent resource that covers key imaging problems such as brain tumors, diabetic retinopathy, breast cancer, fractures, ligament injuries, skin and eye diseases, and neurological disorders, while highlighting the role of deep learning and emerging computational methods.

Each chapter offers an introduction to the medical problem, a review of techniques, and a discussion of challenges and future directions. The book is intended for students, researchers, and professionals who wish to understand both the fundamentals and current trends in intelligent medical imaging.

We thank our coauthors, colleagues, and the publishing team for their contributions and support in making this volume possible.

Shiv Naresh Shivhare, Thipendra P. Singh, Deepa Joshi, and Anitesh Mishra

https://doi.org/10.1515/9783111389059-203

Contents

1 Recent trends, techniques, and algorithms for medical image segmentation and classification

Abstract: Medical image segmentation and classification have become central to computer-aided diagnosis, enabling clinicians to identify pathological regions with precision and efficiency. This chapter surveys recent developments that span imaging modalities, benchmark datasets, and algorithmic strategies ranging from traditional methods to deep learning–based frameworks. Particular attention is given to the way segmentation performance is influenced by modality-specific characteristics, dataset quality, and the choice of evaluation metrics. The discussion also synthesizes challenges that persist in real-world applications, including variability in imaging conditions, class imbalance, and the translation of research prototypes into reliable clinical tools. By examining both the strengths and shortcomings of current practices, the chapter highlights best practices and the countermeasures that have emerged, while outlining promising research directions that may shape the next generation of intelligent medical imaging systems.

Keywords: Medical image segmentation, Computer-aided diagnosis, Biomedical image analysis, Brain tumor, Breast cancer, Diabetic retinopathy, ACL tear, Galucoma, Lung disease, Sking cancer, Alzheimer's disease, Fetal disease

1.1 Introduction

In the field of image processing, image segmentation plays a crucial role in which a digital image is divided into several meaningful regions or segments so that a specific object, such as an organ or tumor, can be identified and its exact location determined. In contrast, image classification classifies the entire image into a single label, such as a cancerous image or noncancerous image. Segmentation not only tells us what is present in the image, but also clearly indicates where it is located. Therefore, segmentation is an essential process in medical image analysis for tasks such as determining the boundaries of abnormal regions or identifying microscopic pathological structures. In recent years, deep learning algorithms have made significant progress, and due to the growing availability of annotated medical datasets and powerful computing hardware, both image segmentation and classification have become popular in medical image analysis. With the help of these techniques, complex patterns in various imaging modalities such as MRI, CT, PET, ultrasound, etc., can be automatically and accurately detected, leading to faster diagnosis, better treatment planning, and improved clinical decision making.

Medical image segmentation is essentially important and highly needed in the domain of medical imaging for various disease diagnosis, early disease detection, survival

https://doi.org/10.1515/9783111389059-001

prediction, and better treatment planning. These advanced imaging modalities have been significantly utilized in the automated detection and segmentation of different chronic and deadly diseases such as brain tumor segmentation, breast cancer detection, bone fracture detection, diabetic retinopathy detection, skin cancer detection, anterior Cruciate Ligament (ACL) detection, Alzheimer's disease detection, etc.

One of the main benefits of medical image segmentation is that it provides physicians and radiologists with precise spatial information that is extremely useful in various critical medical operations such as surgical planning, radiation therapy, and the evaluation of disease progression. It also automatically marks the boundaries of organs and regions affected by the disease, thereby improving both the speed and precision of diagnosis. This is especially important in conditions such as cancer, brain tumors, and eye diseases, where mm-level precision is required. In such cases, segmentation techniques help physicians make lifesaving decisions and are essential to provide appropriate medical consultation.

1.1.1 Image modalities and associated challenges

Segmentation research must constantly adapt to the strengths and weaknesses of the various imaging sources. Instead of treating modalities in isolation, it is helpful to notice their contrasts. For example, MRI is sought when fine-tissue contrast is required, but its sensitivity to motion and lengthy scans often complicate analysis. CT, in comparison, offers speed and clarity for bones, though its radiation dose and weak soft-tissue contrast limit broader adoption. Ultrasound brings the advantages of portability and low cost, but inconsistency across operators and heavy speckle noise lower reliability. X-ray delivers rapid results worldwide, yet the 2D compression of 3D anatomy blurs structural boundaries. PET shifts the perspective by highlighting metabolic processes, though the price paid is high noise, limited resolution, and expense. Retinal fundus images are essential for detecting eye disease, but the fine vascular details and device variability create major hurdles. Taken together, these examples show that segmentation challenges arise not from a single source, but from a spectrum of trade-offs across modalities.

Another way to see these obstacles is thematically rather than modality-by-modality. Some techniques struggle with intensity instability (MRI, ultrasound), others with dimensional compression (X-ray), still others with functional–structural imbalance (PET). Framing the discussion in this comparative way highlights why no universal solution is possible and why modality-aware strategies remain essential. Table 1.1 summarizes the key characteristics of common modalities, highlighting both their clinical value and the obstacles they introduce for automated segmentation.

Table 1.1: Comparative view of major imaging modalities and their segmentation-related challenges.

Modality	Key Strengths	Main Limitations	Segmentation Challenges
MRI	Excellent soft-tissue contrast; useful in neuro- and musculoskeletal studies	Sensitive to motion; long acquisition time; variation in intensity	Nonuniform signals and motion artifacts obscure boundaries; requires normalization and artifact correction
CT	Fast acquisition; sharp bone and dense-structure depiction	Radiation dose; poor soft-tissue contrast	Clear for skeletal features, but ambiguous for soft tissues; requires contrast-enhancement for detail recovery
Ultrasound	Portable, inexpensive, and safe; real-time monitoring	Operator-dependent; speckle noise	Highly variable textures; unstable boundaries; needs noise suppression and robust generalization
X-ray	Quick and low-cost; globally accessible	2D projection of 3D anatomy; overlapping structures	Difficult to isolate boundaries due to compressed layers; contextual separation more important than sharp contours
PET	Provides metabolic/functional insight; valuable in oncology	Low spatial resolution; noisy; expensive	Poor boundary definition, especially for small/diffuse lesions; usually combined with MRI/CT for reliable segmentation
Fundus Imaging	Noninvasive view of retinal structures; crucial for DR and glaucoma	Fine vessels and lesions; variability across devices	High intraclass variation; low contrast in small lesions; requires vessel enhancement and illumination correction

1.1.2 Datasets

Availability of standard and benchmark real datasets plays a crucial role in solving image-based research problems. However, various researchers and organizations have provided various real benchmark datasets for various diseases in the area of medical imaging. A wide range of benchmark datasets has been established to support automated detection, segmentation, and classification tasks in medical imaging research. These datasets cover multiple diseases and imaging modalities, thereby enabling the development and validation of robust computer-aided diagnostic systems.

For musculoskeletal disorders, the ACL-tear dataset from the Rijeka Clinical Hospital in Croatia provides clinical DICOM images that are often used for ligament injury detection. The Center of the Rijeka Clinical Hospital in Croatia provided the standard dataset [359] for the detection of ACL tears. Neuroimaging is strongly represented by resources such as the ADNI dataset, which integrates multimodal MRI, PET, and clinical data for Alzheimer's disease studies, and the OASIS series, which offers MRI data for investigations into aging and dementia. The ADNI (Alzheimer's Disease Neuroimaging

Initiative) [35] dataset is also a popular dataset for the diagnosis of Alzheimer's disease. ADNI offers the multimodal data collected from more than 2,000 participants in significant modalities such as MRI, PET, genetic data, cognitive scores, clinical information, etc. OASIS (Open Access Series of Imaging Studies) [79] is another benchmark dataset for the investigation of brain aging and Alzheimer's disease. In the different versions of OASIS, different data modalities such as T1-weighted MRI, clinical data, demographics, segmentation labels have been provided for automated problem-solving.

Similarly, brain-tumor research has been advanced by the BRATS challenge datasets, which include multicontrast MRI scans, as well as complementary repositories like TCIA and BrainWeb. Brain-tumor detection and segmentation is another popular research problem of medical imaging domain. BRATS [177] have been a benchmark dataset from 2012 onward in different versions containing MRI scans in four image modalities such as T1-weighted, T2-weighted, contrast-enhanced T1-weighted and FLAIR. However, there are also several other significant sources of datasets for brain imaging research such as TCIA [243] and BrainWeb [59]. In oncology, breast-cancer analysis is supported by large-scale histopathology datasets, including the Digital Pathology dataset and Databiox, both of which provide high-resolution tissue slides. The digital pathology [118] and Databiox [44] datasets provide high-resolution histopathological slides scans for diagnosis, prognosis, and AI-driven insights for the detection of breast cancer.

Retinal imaging has also received significant attention, particularly for diseases such as diabetic retinopathy and glaucoma. The IDRiD and DIARETDB1 datasets include annotated fundus images for retinopathy classification, while DRISHTI-GS, RIM-ONE, and REFUGE provide challenging benchmarks for glaucoma detection. For fetal health monitoring, the Fetus Phantom dataset offers ultrasound images capturing fetal orientation and anatomical features. Indian Diabetic Retinopathy Image Dataset (IDRiD) [212] and DIARETDB1 [127] were prepared to support the classification of diseases for diabetic retinopathy, having numerous annotated color fundus image scans. The phantom fetus dataset [214] contains ultrasound images in the form of fetus orientation, fetal planes, anatomical features, etc. Glaucoma disease can damage the optic nerve and may lead to vision loss. DRISHTI-GH [265], RIM-ONE [80], and REFUSE [229] are existing challenging data sets with images of the retinal fundus to identify characteristic changes associated with glaucoma. Prostate cancer imaging is represented by MRI datasets curated by Brigham and Women's Hospital, while skin-cancer detection research often relies on dermoscopic repositories such as ISIC and PH2. Pulmonary and thoracic imaging is supported by datasets like the NIH Chest X-ray collection, widely used for lung-disease classification, and the ACDC-LungHP histopathology dataset designed for lung-cancer studies. Together, these repositories not only cover a broad spectrum of diseases but also provide multimodal data-ranging from MRI and CT to ultrasound, X-ray, and histopathology-ensuring comprehensive support for developing AI-driven diagnostic methods. Table 1.2 presents a brief description and the availability of various benchmark datasets for the automated detection and segmentation of various serious diseases, including the source of the dataset. Figure 1.1 presents the sample medical images from various datasets.

Table 1.2: Datasets description and availability for automated detection and segmentation of various severe diseases.

Disease Type	Dataset	Image Modality	Dataset Source
ACL	[359]	DICOM images	Clinical Hospital Centre Rijeka, Croatia
Alzheimer's disease	ADNI [35]	T1, T2 MRI	https://adni.loni.usc.edu/
Alzheimer's disease	OASIS [79]	T1 MRI	https://www.oasis-brains.org/
Brain tumor	BRATS [177]	T1, T2, T1C, Flair	https://www.kaggle.com/datasets/awsaf49/brats2020-training-data
Brain cancer	TCIA [243]	MRI	https://www.cancerimagingarchive.net/
Brain cancer	BrainWeb [59]	T1, T2, and Proton-density	http://www.bic.mni.mcgill.ca/brainweb/
Breast cancer	Digital pathology [118]	Histopathology images	https://www.kaggle.com/datasets/paultimothymooney/breast-histopathology-images
Breast cancer	Databiox [44]	Histopathology images	https://databiox.com/
Diabetic retinopathy	IDRiD [212]	Retinal images	https://idrid.grand-challenge.org/
Diabetic retinopathy	DIARETDB1 [127]	Retinal images	https://www.kaggle.com/datasets/nguyenhung1903/diaretdb1-v21
Fetal disease	Fetus Phantom [214]	Ultrasound images	https://github.com/bharathprabakaran/FPUS23
Glaucoma	DRISHTI-GS [265]	Eye fundus images	https://www.kaggle.com/datasets/lokeshsaipureddi/drishtigs-retina-dataset-for-onh-segmentation
Glaucoma	RIM-ONE v3 [80]	Eye fundus images	https://www.idiap.ch/software/bob/docs/bob/bob.db.rimoner3/stable/index.html
Glaucoma	REFUGE-2020 [229]	Eye fundus images	https://refuge.grand-challenge.org/iChallenge-AMD/
Lung disease	NIH Chest [317]	X-ray images	https://www.kaggle.com/datasets/nih-chest-xrays/data
Lung cancer	ACDC-LungHP [150]	Histopathology images	https://acdc-lunghp.grand-challenge.org/
Prostate cancer	The Brigham and Women's Hospital	MRI	https://prostatemrimagedatabase.com/index.html
Skin cancer	ISIC-2020 [233]	DICOM images	https://challenge.isic-archive.com/data/#2020
Skin cancer	PH2 [175]	RGB images	https://www.fc.up.pt/addi/ph2%20database.html

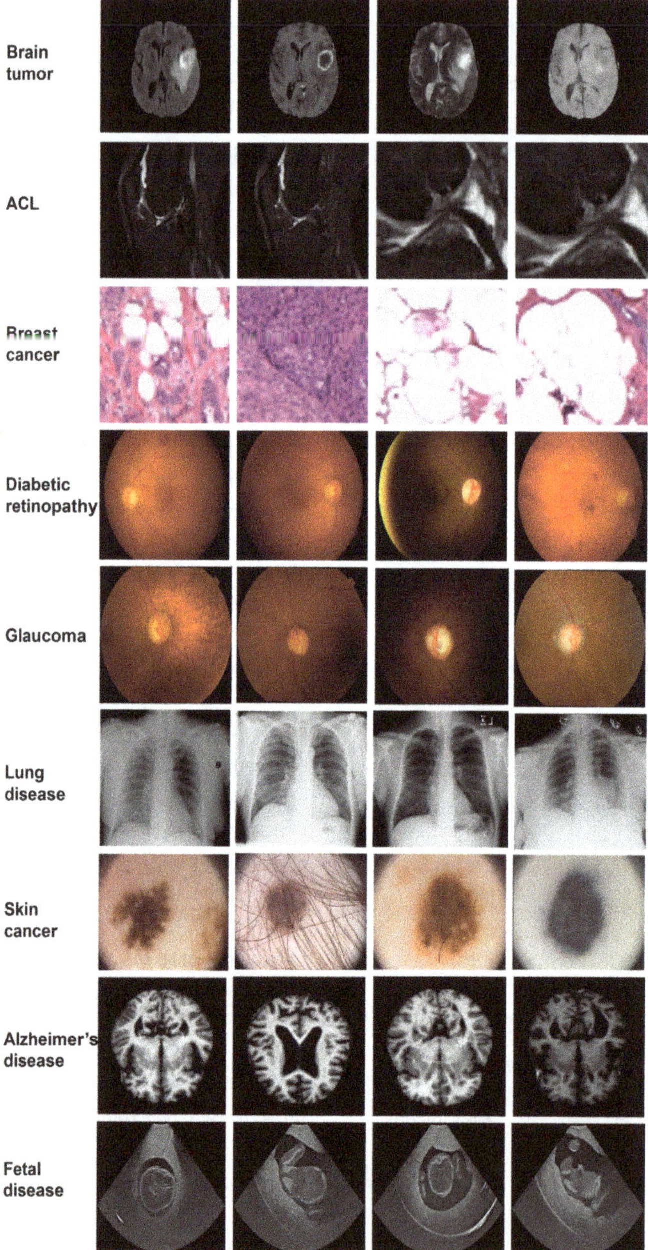

Figure 1.1: The sample images from various datasets for the automated disease detection and segmentation. [35, 44, 59, 79, 80, 118, 127, 150, 175, 177, 212, 214, 229, 265].

1.1.3 Performance metrics for medical image segmentation

Quantitative evaluation of medical image segmentation methods involves various performance metrics such as the Dice Similarity Coefficient (DSC), Positive Predictive Value (PPV), Sensitivity, Accuracy, Specificity, Jaccard Index, along with metrics like the Hausdorff distance and computation time. The DSC assesses the agreement between the predicted and actual results by measuring the overlap. It is calculated as the harmonic mean of the PPV and sensitivity. Sensitivity measures the correct identification of positive instances, while PPV assesses the ratio of true positive predictions to all positive predictions. Accuracy quantifies all correct predictions, including both true positives and true negatives. Specificity indicates the rate of true negative predictions among all negative predictions. The Jaccard index [37] measures the intersection of the predicted and actual regions. The Hausdorff distance [110] compares the predicted segmentation with the ground truth, assessing the maximum error between the corresponding points. It is defined as the maximum distance between the points in one set and the nearest point in the other set as shown in Eq. (1.7). These metrics provide insights into the effectiveness of image-segmentation techniques. The mathematical formulations of these performance measures are as follows:

$$\text{Dice Similarity Coefficient} = \frac{2 \times \text{TP}}{(\text{TP} + \text{FP}) + (\text{TP} + \text{FN})} \tag{1.1}$$

$$\text{Sensitivity} = \frac{\text{TP}}{\text{TP} + \text{FN}} \tag{1.2}$$

$$\text{Positive Predictive Value} = \frac{\text{TP}}{\text{TP} + \text{FP}} \tag{1.3}$$

$$\text{Accuracy} = \frac{\text{TP} + \text{TN}}{\text{TP} + \text{TN} + \text{FP} + \text{FN}} \tag{1.4}$$

$$\text{Specificity} = \frac{\text{TN}}{\text{TN} + \text{FP}} \tag{1.5}$$

$$\text{Jaccard Index} = \frac{\text{TP}}{\text{TP} + \text{FP} + \text{FN}} \tag{1.6}$$

$$h(X, Y) = \max_{x \in X}\{\min_{y \in Y}\{d(x, y)\}\} \tag{1.7}$$

True Positive (TP) denotes the precise identification of a tumor, while True Negative (TN) signifies the accurate identification of a nontumor. False Positive (FP) instances arise when the prediction indicates a tumor, contrary to the actual nontumor ground truth, reflecting inaccuracies in prediction. Conversely, False Negative (FN) occurrences denote incorrect predictions where the model suggests a nontumor, despite the ground truth being a tumor. Among the specified metrics, the Dice Similarity Coefficient (DSC) emerges as the most critical and widely utilized tool for evaluating brain-tumor segmentation methods. It offers consensus by quantifying the overlap between the predicted and actual regions.

1.2 Medical image segmentation and analysis algorithms

Medical image segmentation has been explored through a range of traditional machine learning and modern deep-learning approaches. Earlier frameworks often relied on handcrafted features, statistical models, or conventional classifiers such as k-nearest neighbor and support vector machines. While these approaches offered interpretability and were computationally less demanding, their performance was highly dependent on carefully designed features, and they often struggled with variability across imaging modalities or patient populations. Deep-learning models, in contrast, learn hierarchical features directly from data, which reduces reliance on manual feature extraction. Yet, they typically require large amounts of annotated training data and significant computational resources. A recurring challenge across both paradigms is the trade-off between accuracy, generalization, and efficiency.

The methodologies reported in the literature differ widely, ranging from classical CNN-based models and transfer-learning pipelines to cascaded architectures, ensemble methods, and attention-enhanced networks. Some techniques emphasize robustness against noisy or limited datasets, while others focus on computational efficiency by reducing parameter counts. A few approaches also integrate contextual information or multi-scale pathways to better capture the complex nature of medical images. Each of these methodological choices reflects the need to balance precision, scalability, and clinical applicability. Deep learning has become the dominant paradigm in medical image analysis, particularly for segmentation tasks. Unlike traditional machine learning, where features are engineered manually, deep-learning models automatically extract low- to high-level representations from raw data [249]. This shift makes possible the capture of subtle patterns and nonlinear structures that handcrafted methods often miss. Architectures such as U-Net, ResNet, and GAN-based models have demonstrated superior performance across imaging modalities including MRI, CT, ultrasound, histopathology, and fundus photography. The key distinction lies in representation learning: deep networks not only outperform traditional models in accuracy but also adapt more effectively to multimodal datasets. However, their reliance on large, well-annotated data and their "black-box" nature remain significant limitations for clinical translation.

Chang et al. [47] proposed a CNN integrated with residual blocks for detecting ACL tears. Their approach used a dynamic patch-based sampling strategy, which improved model accuracy. However, its generalization capability declined when applied to larger datasets. Key et al. [129] applied a k-NN model that incorporated deep features extracted from AlexNet. The use of the iterative ReliefF algorithm helped refine feature selection and boosted classification results. Even so, the method's success strongly relied on the quality of the pre-trained model. Liu et al. [155] designed a cascaded CNN that performed section detection, ligament isolation, and classification in sequential stages. This modular approach enhanced the detection process. The downside, however, was the long computational time required when working with smaller datasets.

Ebrahimi et al. [72] employed a CNN configured with ResNet-18 pretrained on ImageNet to analyze MRI scans. The model achieved improved accuracy in Alzheimer's detection, yet it struggled to adapt when applied to purely 2D data representations. Folego et al. [78] tested several CNN variants including LeNet-5, VGG, GoogLeNet, and ResNet. Their pipeline offered a fully automated and computationally efficient solution, but its accuracy was still limited compared to more advanced designs. Ghazal et al. [85] utilized a transfer learning approach with AlexNet, thereby removing the dependence on manually engineered features. Although this streamlined the feature extraction process, the overall outcomes were strongly tied to the pretrained network's capability. Liu et al. [156] introduced a depthwise separable CNN, which reduced parameter count and computational load. This efficiency gain came at the cost of greater risks of overfitting or underfitting, particularly on small-scale datasets. In a follow-up, Liu et al. [158] also experimented with a 3D CNN that leveraged region-of-interest (ROI) features for volumetric analysis. This approach successfully improved feature representation but demanded extensive computational resources, making it less practical for routine use.

Orlando et al. [200] customized the U-Net architecture by integrating a 3D reconstruction module to process prostate MRI scans. This modification allowed the model to retain spatial and structural information across slices, thereby improving segmentation accuracy. However, the reliance on fixed step angles for slice generation sometimes limited performance consistency and added computational overhead.

Huang et al. [108] introduced an intensity-invariant deep neural network that incorporated an adaptive gamma correction block. This design improved robustness against variations in image intensity. Still, its applicability was restricted mainly to glioma segmentation. Majie et al. [167] developed an Attention Res-UNet with a guided decoder. The guided decoding process enhanced the extraction of discriminative features, yet the model underperformed when segmenting tumor cores and enhancing tumor regions. Sun et al. [276] designed a 3D fully convolutional framework with a multi-pathway architecture. The parallel pathways helped capture diverse features from MRI scans, but the approach struggled to accurately segment tumor cores and enhancing tissues. Ye et al. [332] proposed a dual-path network that processed contextual and detailed cues simultaneously. By combining these two perspectives, the model achieved improved accuracy, although it required significant training time. Zhang et al. [342] employed an ensemble of three U-Net models, each focusing on feature extraction at different stages. The combined outputs boosted segmentation quality, but the training of multiple U-Nets demanded large memory resources and longer computation time. Zhou et al. [353] applied atrous convolutional layers within a CNN for semantic segmentation. The dilation strategy improved boundary recognition, but performance dropped when handling smaller tumor regions.

Ilesnamie et al. [112] implemented a VE block–based U-Net model on enhanced ultrasound images of breast cancer. By combining max and average pooling operations, the network was able to capture important low-level features. However, performance could be further improved by integrating more advanced deep-learning modules. Van et al. [309] employed a CNN-driven semantic segmentation approach that merged contextual and resolution-based details for more accurate lesion detection. While effective, the approach required careful tuning of control parameters to reach optimal results. Yan et al. [330] focused on mass detection and segmentation from X-ray scans. Their multiscale, batch-based segmentation framework successfully identified multiple lesions from a single scan. Nonetheless, annotation noise and variability still posed challenges. Zou et al. [357] developed a noise-index–based method to detect and correct mislabeled data during backpropagation. This strategy helped in handling noisy datasets but remained dependent on limited labeled data and incurred high annotation costs.

Arsalan et al. [17] proposed a dual residual-stream network for retinal-vessel segmentation of diabetic retinopathy. The architecture delivered high accuracy with fewer parameters, but the feature maps were sometimes insufficient to capture very fine vessel details. Qureshi et al. [221] introduced an active deep-learning system with a multilayer setup that minimized the need for large labeled datasets. While label efficiency was a major advantage, performance tended to decline when scaling to larger datasets. Sambyal et al. [240] designed a modified U-Net that incorporated ResNet layers and periodic shuffling. This hybrid architecture significantly improved segmentation accuracy, although it lacked generalizability due to the absence of clinically acquired data. Skouta et al. [266] created a modified U-Net model aimed at early diagnosis of diabetic retinopathy. Their framework enhanced the detection of retinal hemorrhages, but it was limited in its ability to accurately grade disease severity.

Chutia et al. [56] applied a pretrained DenseNet-201 model that was modified with global average pooling and an added dense layer. This adaptation improved precision, recall, and F1-score for lung disease classification. However, its performance was not consistent across other datasets. Sridevi and Kannan [272] introduced a 3D TransDense U-Net++ with a novel loss function for lung-cancer detection. The model achieved high precision and accuracy, supporting early diagnosis. Its drawback was the heavy computational cost associated with training. Talib et al. [287] combined transformer layers with deep learning using patch embeddings. This setup allowed for fine-grained feature extraction and improved segmentation performance. Nonetheless, the method needed better preprocessing strategies to avoid image quality trade-offs. Wang et al. [313] developed a federated semi-supervised learning framework for lung-tumor segmentation. The approach effectively addressed privacy issues and the scarcity of annotations. Still, it struggled to handle heterogeneous data distributions across different institutions.

Bian et al. [40] proposed a modified U-Net enhanced with a GAN framework that removed the need for separate pre- and post-processing steps. However, errors produced in the first stage could negatively influence subsequent segmentation outcomes. Imtiaz et al. [113] employed transfer learning for semantic segmentation of optic disc and cup regions. The approach was efficient and required less training time, but distinguishing closely located structures such as the optic disc and cup remained a challenge. Morano et al. [181] worked on segmentation of retinal vessels into arteries, veins, and smaller branches using a neural network. This method produced detailed vessel maps but occasionally suffered from inconsistencies in vessel classification. Pachade et al. [203] utilized a Nested EfficientNet model that required fewer computational resources compared to traditional CNNs. Although resource-friendly, the framework struggled to deliver consistent results across diverse datasets. Sun et al. [278] integrated contextual information into a U-Net with dilated convolution modules, leading to improved performance. Yet, the model was computationally expensive to train. Tulsani et al. [302] created a customized U-Net trained with binary cross-entropy and Dice coefficient losses to address class imbalance. Despite reducing over- and under-segmentation, the method still faced difficulty with capillary-level details. Yin et al. [333] applied an active contour model powered by a deep neural network. This approach maintained region consistency and smooth boundaries, though it relied heavily on the underlying network backbone. Yuan et al. [336] cascaded two U-Nets (W-Net) to extract deeper semantic and structural features. The design improved segmentation quality but had challenges in localizing optic disc centers in low-quality images. Zhang et al. [346] adapted U-Net with attention modules to capture high-level domain-shift–resistant features. However, the attention layers were sensitive to domain variations, causing reduced cross-dataset accuracy. Zhou et al. [352] explored a GAN-based U-Net, where the architecture acted as both generator and discriminator. This design achieved strong results across multiple datasets but sometimes produced false positives along vessel boundaries due to its focus on local details.

Arora et al. [16] adapted a U-Net architecture by replacing batch normalization with group normalization and adding atrous convolutions. This combination improved feature propagation and boosted segmentation performance, though it remained dependent on careful pre-and post-processing. Hasan et al. [96] designed a depthwise separable CNN to reduce the parameter count and improve handling of regions of interest. While efficient, the absence of pooling layers sometimes resulted in weaker feature resolution. Jin et al. [124] proposed a hybrid framework that combined rooted segmentation with a cascade knowledge-diffusion network. The modular design supported faster convergence and reliable deployment, but the training process was comparatively complex. Gu et al. [88] applied a ResNet-50 backbone with an edge-guided module to capture lesion boundary details. Although this approach strengthened edge detection, variability in lesion borders occasionally reduced segmentation reliability. Sarker et al. [241] introduced a lightweight GAN-based model that produced strong results across multiple public datasets. Despite this robustness, performance decreased when lesions were positioned near image margins or lacked clear boundaries. Tang et al. [289] presented an ensemble model founded on adaptive feature learning. The method proved effective in handling low-contrast lesions, but it struggled when segmenting lesions with extremely poor visibility. Zhang et al. [343] worked on additional ensemble-based segmentation frameworks that leveraged multi-stage U-Nets, further enhancing segmentation accuracy. Nevertheless, the large memory footprint and high training cost limited scalability.

Pu et al. [217] developed a hybrid framework that combined convolutional and recurrent neural networks, further enhanced by dual feature-fusion strategies. This design effectively captured both static spatial and dynamic motion-related features from ultrasound data. Its limitation, however, was susceptibility to lag and uncertainty in modeling contextual information. Prabakaran et al. [214] created the FPUS23 dataset and trained a ResNet-34–based model for fetal-feature identification. The system achieved high accuracy with minimal fine-tuning, but it occasionally misclassified artifacts that appeared during probe navigation, particularly when fetal anatomies were complex.

From the surveyed methods, it is evident that deep learning has become the central driver of progress in medical image segmentation. These models are characterized by their ability to automatically learn discriminative features from large volumes of imaging data, eliminating the need for handcrafted descriptors. Architectures such as U-Net and its numerous variants are particularly well-suited to medical tasks, as their encoder–decoder structure preserves both global context and fine structural details. Extensions using attention modules, transfer learning, or multi-pathway designs further enhance adaptability across different imaging modalities. Table 1.3 presents the summary of recent research works for medical images detection and segmentation.

The primary advantages of deep learning models include their superior accuracy, capacity to generalize to complex datasets, and robustness in handling multimodal inputs such as MRI, CT, and histopathology slides. They are also scalable, allowing integration of large-scale data for disease-specific benchmarks. On the downside, these methods are data-hungry, often requiring extensive annotated datasets for training. Computational demands are another limitation sinces many networks involve millions of parameters and require GPU support for practical deployment. Moreover, the "black box" nature of deep models raises concerns regarding interpretability and clinical trust, making validation and regulatory acceptance more challenging. Overall, deep learning has shifted the landscape of medical image segmentation by moving beyond feature engineering to fully data-driven representation learning. Its advantages far outweigh traditional approaches in terms of performance, but future research must continue to address the issues of explainability, efficiency, and reliance on large-scale labeled datasets.

1.3 Performance analysis and discussion

Table 1.4 aggregates Dice/Jaccard (overlap), Sensitivity/PPV/Specificity (recognition quality), Accuracy/AUC (overall discrimination), and the Hausdorff Distance (boundary fidelity) across public and private datasets. Scores vary widely by modality, task granularity (region vs whole-organ), and dataset difficulty, so inter-row comparisons

Table 1.3: Summary of the recent research works for medical images detection and segmentation. CNN: Convolutional Neural Network, kNN: k-Nearest Neighbor.

Disease Type	Author(s)	Methodology	Advantages	Limitations
ACL	Chang et al. [47]	CNN with Residual block	Dynamic patch-based sampling improves performance	Performance may vary in large dataset
	Key et al. [129]	kNN with AlexNet-based features	Deep features selected by iterative RelifF algorithm improves performance	Performance depends on the pretrained model
	Liu et al. [155]	Cascaded CNN	Effectively performs section detection, ligament isolation and classification separately.	Large training time on small dataset.
Alzheimer's disease	Ebrahimi et al. [72]	CNN with ResNet-18 pre-trained on ImageNet	improved classification accuracy	Sequence-based model may struggle on 2D datasets.
	Folego et al. [78]	CNN with LeNet-5, VGG, GoogLeNet, and ResNet	Fully automated and efficient	Limited performance
	Ghazal et al. [85]	Transfer learning with AlexNet model	Does not require hand-crafted features	Performance depends on the pre-trained model.
	Liu et al. [156]	Depthwise separable convolution neural network	Reduced parameters and computational costs	May lead to overfitting or underfitting in the case of small dataset.
	Liu et al. [158]	3D CNN	Deep learning and ROI-based features improve performance	Computationally complex
Prostate cancer	Orlando et al. [200]	Modified U-Net with 3D reconstruction	3D image reconstruction preserved spatial and structural information	Use of fixed step angle for slice generation may impact performance and computation time

Table 1.3 (continued)

Disease Type	Author(s)	Methodology	Advantages	Limitations
Brain tumor	Huang et al. [108]	Intensity-invariance deep neural network	adaptive gamma correction block performs intensity invariance	Limited to segments brain tumors with gliomas only.
	Maji et al. [167]	Attention Res-UNet with Guided Decoder	Guided decoder helps in extracting improved features	Performance lags for tumor core and enhancing tumor regions.
	Sun et al. [276]	3D fully convolutional network	Multi-pathway architecture extracts effective features	Performance lags for tumor core and enhancing tumor regions.
	Ye et al. [332]	Parallel architecture of attention and context pathways	Model utilizes both contextual and detail cues of MRI scans	Large training time
	Zhang et al. [342]	Combination of three U-Nets	Extracting significant features from multiple U-Nets at different stages enhances the performance	Training multiple U-Nets requires large memory and training time.
	Zhou et al. [353]	Atrous convolution based CNN	Semantic segmentation performs better with atrous convolution	Performance on small-size brain tumors needs improvement.
Breast cancer	Ilesnami et al. [112]	VE block-based U-Net model on enhanced ultrasound images	Combination of max and average pooling extracts significant low-level features	Model could be enhanced using advanced deep learning methods
	Van et al. [309]	CNN-based semantic image segmentation	Model combines contextual and resolution details for accurate segmentation	The value of control parameter could be tuned to optimize performance
	Yan et al. [330]	Mass detection and segmentation from X-ray scans	Can detect multiple masses from one X-ray scan	Multiscale batch-based segmentation improves performance
	Zou et al. [357]	Detects and corrects noisy labels of the dataset using noise index	Helpful to handle the noisy data while backpropagation learning	Dependence on limited data, High cost of manual annotations

Table 1.3 (continued)

Disease Type	Author(s)	Methodology	Advantages	Limitations
Diabetic Retinopathy	Arsalan et al. [17]	Dual-residual-stream-based network for retinal vessel segmentation	High and efficient segmentation accuracy with reduced complexity and fewer parameters	Feature map size can be enhanced to capture finer details
	Qureshi et al. [221]	Active deep learning system with multilayer architecture	Label-efficient active learning reduces the need for extensive labeled datasets	Performance may face challenges with large-scale dataset
	Sambyal et al. [240]	Modified U-Net architecture incorporating a resnet and periodic shuffling	High performance	Limited generalizability due to lack of clinically acquired data
	Skouta et al. [266]	Modified U-Net architecture aiding in early diagnosis of diabetic retinopathy	Enhances the detection accuracy of retinal hemorrhages	Could be enhanced to grade diabetic retinopathy
Lung disease	Chutia et al. [56]	Pretrained DenseNet201 model by adding global average pooling and a dense layer	Demonstrates enhanced precision, recall, and F1-score	Performance may vary in other datasets
	Sridevi and RajivKannan [272]	Use of 3D Trans-DenseUnet++ with a new loss function	High accuracy and precision demonstrating effective early diagnosis of lung cancer	High computational cost
	Talib et al. [287]	Transformer and deep learning	Patch-embedding layers ensure fine-grained feature extraction for improved performance	Need for improved preprocessing methods to better adjust image without compromising quality
	Wang et al. [313]	Federated semi-supervised learning	Addresses data privacy issues and lack of annotated data in lung-tumor segmentation	Inability to effectively handle data heterogeneity at the algorithmic level

Table 1.3 (continued)

Disease Type	Author(s)	Methodology	Advantages	Limitations
Glaucoma	Bian et al. [40]	Modified U-Net with GAN architecture	Preprocessing and post-processing is not needed	Performance of second stage can be negatively impacted if the first stage produces inaccurate results
	Imtiaz et al. [113]	Transfer learning-based semantic segmentation	Low training time	Semantic segmentation may suffer to distinguish between closely located instances of OD and OC
	Morano et al. [181]	Segmentation of retinal vessels into arteries, veins, and vessels using neural network	Simultaneously provides detailed segmentation maps for arteries, veins, and all vessels	The approach may suffer from vessel classification incoherence and relies on low-level features
	Pachade et al. [203]	Nested EfficientNet	Requires less computational resources	May suffer to produce optimized results in different datasets
	Sun et al. [278]	Contextual information enabled U-net	Model with dilated convolutional modules enhances performance	Computationally expensive
	Tulsani et al. [302]	Customized U-Net	Binary cross-entropy and dice coefficient solves class imbalance	Over and under segmentation due to blood capillaries
	Yin et al. [333]	Deep neural network enabled active contour model	Works well with region consistency and boundary smoothness	its reliance on a specific network backbone
	Yuan et al. [336]	Cascading two U-Nets as W-Net	Extracts deeper structural and semantic information	Challenges in OD center localization in low-quality images
	Zhang et al. [346]	Modified U-Net with attention modules that extracts high-level features	handles domain shift in Optical Disc and Cup segmentation	Sensitivity of attention modules to domain shifts, leading to inferior result in cross-dataset tasks
	Zhou et al. [352]	U-Net as both generator and discriminator in GAN	Achieves superior results across multiple datasets	More focus on local details leads to false positives at vessel edges

Table 1.3 (continued)

Disease Type	Author(s)	Methodology	Advantages	Limitations
Skin cancer	Arora et al. [16]	Modified U-Net architecture with group normalization instead of batch normalization	Atrous convolutions enhanced feature propagation achieving superior performance	Depends on preprocessing and post-processing
	Hasan et al. [96]	Depth-wise separable convolution to reduce parameters	Enhanced performance by better handling ROI selection	Absence of pooling layers may lead to poor feature resolution
	Jin et al. [124]	Integration of Rooted Segmentation and cascade knowledge-diffusion network	Balanced architecture and modular design ensure faster convergence and effective deployment	complexity of the training procedure
	Gu et al. [88]	ResNet-50 model with five convolutional blocks	Edge information-guided module extracts boundary details	Method may struggle with variability in lesion boundaries
	Sarker et al. [241]	Lightweight GAN-based model	Strong results across multiple datasets	Performance lags when skin lesions intersect image margins or lack clear boundaries
	Tang et al. [289]	Ensemble model based on adaptive feature learning	Robustness in handling low-contrast lesions	Method struggles with segmenting lesions having low contrast
	Zhang et al. [343]			–
Fetal disease	Pu et al. [217]	Combination of convolutional and recurrent neural networks, enhanced by two feature fusion strategies	Ability to effectively capture both static spatial features and dynamic motion information	Performance may affect due to lag and uncertainty in capturing context information
	Prabakaran et al. [214]	Creates the FPUS23 dataset, Trains ResNet34-based model for fetal features	Ability to achieve high accuracy with minimal fine-tuning	Misclassifies certain abstract artifacts during probe navigation as fetal anatomies

Table 1.4: Performance of recent research works for medical images detection and segmentation in terms of Dice Similarity Coefficient (DSC), Sensitivity, Positive Predictive Value (PPV) AC: Accuracy, JI: Jaccard Index, SP: Specificity, ACL: Anterior cruciate ligament, AUC: Area Under the Curve, IoU: Intersection of Union, BUS: Breast Ultrasound Images, HD: Hausdorff Distance.

Disease Type	Author (s)	Dataset	DSC	Sensitivity	PPV	Other Score
ACL	Chang et al. [47]	Private	–	0.79	0.75	0.75 (AC)
	Key et al. [129]	Private	0.92	0.95	0.90	0.98 (AC)
	Liu et al. [155]	Private	–	0.92	–	0.92 (SP)
Alzheimer's disease	Ebrahimi et al. [72]	ADNI [35]	–	0.91	–	0.91 (AC)
	Folego et al. [78]	ADNI [35]	–	–	–	0.52 (AC)
	Ghazal et al. [85]	Kaggle [71]	–	–	–	0.91 (AC)
	Liu et al. [156]	OASIS [79]	–	–	–	0.87 (AC)
	Liu et al. [158]	ADNI [35]	–	–	–	0.85 (AUC)
Brain tumor	Huang et al. [108]	TCIA [243]	0.85	0.87	–	0.80 (IoU)
	Maji et al. [167]	BRATS 2019 [177]	0.91	–	–	0.83 (IoU)
	Sun et al. [276]	BRATS 2019 [177]	0.89	0.88	–	0.99 (SP)
	Ye et al. [332]	BRATS 2017 [177]	0.88	0.84	–	–
	Zhang et al. [342]	BRATS 2015 [177]	0.85	–	–	–
	Zhou et al. [353]	BRATS 2015 [177]	0.81	–	–	–
Breast cancer	Ilesnami et al. [112]	BUS [231]	0.90	–	–	7.8 (HD)
	Van et al. [309]	Private	0.87	–	–	–
	Yan et al. [330]	INbreast [182]	0.90	–	–	–
	Zou et al. [357]	Private	0.87	0.88	0.87	0.79 (IoU)
Diabetic retinopathy	Arsalan et al. [17]	DRIVE [273]	–	0.80	–	0.96 (AC)
	Qureshi et al. [221]	EyePACS [55]	0.93	0.92	–	0.95 (SP)
	Sambyal et al. [240]	IDRiD [212]	0.99	0.99	–	0.99 (AC)
	Skouta et al. [266]	IDRiD [212]	0.86	0.80	0.99	0.98 (AC)

Table 1.4 (continued)

Disease Type	Author (s)	Dataset	DSC	Sensitivity	PPV	Other Score
Glaucoma	Bian et al. [40]	REFUGE-2020 [229]	0.93 (Optic Disc)	–	–	8.9 (HD)
	Imtiaz et al. [113]	Drishti-GS [265]	0.94 (Optic Disc)	0.96	–	0.99 (AC)
	Morano et al. [181]	DRIVE [273]	–	0.76 (Vessels)	–	0.95 (AC)
	Pachade et al. [203]	REFUGE-2020 [229]	0.96 (Optic Disc)	–	–	–
	Sun et al. [278]	DRIVE [273]	0.82 (Vessels)	0.79	–	0.96 (AC)
	Tulsani et al. [302]	Drishti-GS [265]	0.96 (Optic Disc)	–	–	–
	Yin et al. [333]	Drishti-GS [265]	0.96 (Optic Disc)	–	–	–
	Yuan et al. [336]	Drishti-GS [265]	0.97 (Optic Disc)	0.95	–	0.98 (AC)
	Zhang et al. [346]	Drishti-GS [265]	0.96 (Optic Disc)	–	–	–
	Zhou et al. [352]	DRIVE [273]	0.83 (Vessels)	0.82	–	0.95 (AC)
Lung disease	Chutia et al. [56]	Chest X-ray [317]	0.96	0.98	0.96	0.96 (AUC)
	Sridevi and RajivKannan [272]	CT-scans [93]	0.86	0.93	0.81	0.92 (AC)
	Talib et al. [287]	ACDC-LungHP [150]	0.52	–	–	0.49 (IoU)
	Wang et al. [313]	CT-scans [258]	0.84	–	–	–
Prostate cancer	Orlando et al. [200]	Private	0.94	0.96	0.93	2.8 (HD)
Skin disease	Arora et al. [16]	ISIC-2018 [58]	0.91	0.94	–	0.95 (AC)
	Hasan et al. [96]	PH2 [175]	–	0.92	–	0.87 (IoU)
	Jin et al. [124]	ISIC-2017 [58]	–	0.70	0.73	0.90 (AC)
	Gu et al. [88]	ISIC-2017 [58]	0.87	0.88	–	0.94 (AC)
	Sarker et al. [241]	ISIC-2017 [58]	0.90	0.87	–	0.97 (AC)
	Tang et al. [289]	ISIC-2017 [58]	0.88	–	–	0.94 (AC)
Fetal disease	Pu et al. [217]	Private	0.85	0.85	0.85	0.85 (AC)
	Prabakaran et al. [214]	Fetus Phantom [214]	–	–	–	0.92 (AC)

should be read with caution. The disease-wise major analysis and observations are summarized as follows.

- *ACL*: Recent pipelines on private cohorts report strong overall accuracy (≈ .98) and balanced Sensitivity/PPV around 0.9, reflecting reliable ligament detection when evaluation settings match training data.
- *Alzheimer's disease*: MRI classifiers on ADNI/OASIS commonly report Accuracy ≈ .85–0.91 (and AUC ≈ .85), indicating dependable screening performance but still short of segmentation-level overlap metrics seen in other tasks.
- *Brain tumor*: On BRATS/TCIA, Dice ≈ .81–0.91 is typical; gains appear for whole-tumor masks, while tumor core/enhancing subregions lag, mirroring the difficulty of small.
- *Breast cancer*: BUS and mammo/histo studies show Dice ≈ .87–0.90 with HD around 7–8 px, suggesting good lesion localization but with boundary roughness remaining a limiter; noise-handling frameworks modestly lift IoU (≈ .79).
- *Diabetic retinopathy*: Optic-disc/lesion pipelines on IDRiD/DRIVE/EyePACS achieve near-ceiling Accuracy (≈ .98–0.99) and Dice up to 0.99, whereas vessel maps are harder (Dice ≈ .82–0.86).
- *Glaucoma*: Optic disc Dice ≈ .93–0.97 (DRISHTI-GS/REFUGE) and HD ≈ −9px show robust disc delineation; vessel segmentation again trends lower (Dice ≈ .83).
- *Lung disease/cancer*: Chest X-ray settings report high marks (Dice ≈ .96; AUC ≈ .96), while CT tumor segmentation is tougher (Dice ≈ .84–0.86) and may drop on challenge sets (e. g., IoU ≈ .49 on ACDC-LungHP), underscoring 3D/contrast variability.
- *Prostate cancer*: On private MRI cohorts, Dice ≈ .94 with HD 2.8 px highlights very clean boundaries, though external generalization isn't directly verifiable.
- *Skin disease*: ISIC/PH2 studies report Dice ≈ .87–0.91 and Accuracy ≈ .94–0.97; boundary-sensitive metrics (IoU/HD) reveal drop-offs for low-contrast, margin-touching lesions.
- *Fetal disease*: Ultrasound pipelines typically show DSC/ACC ≈ .85–0.92, with performance impacted by probe artifacts and anatomical variability.

1.3.1 Major observations

- *Task difficulty is disease- and region-dependent.* Optic disc and whole-tumor masks score higher than vessel or tumor-core regions, reflecting sensitivity to fine structures and class imbalance.
- *Modality matters.* X-ray tasks often attain higher headline metrics than CT/MRI segmentation of small or diffuse lesions, while histopathology shows strong overlap but noisier boundaries (HD).
- *Metric choice shapes the story.* High Accuracy/AUC doesn't guarantee tight boundaries (HD/IoU), and Dice can obscure small-structure errors; reporting multiple metrics is essential for fair comparison.

- *Data setting drives scores.* Results on private sets and well-curated challenges are not directly comparable; domain shift can erode Dice/IoU even when accuracy remains high.
- *Deep learning helps—but costs compute.* Attention/multi-path/3D models push Dice upward on BRATS/retina tasks, while label-efficiency (active learning/federated) mitigates annotation limits yet may underperform on heterogeneous multisite data.

1.4 Challenges in medical image segmentation

Although medical imaging modalities provide critical diagnostic information, each introduces its own obstacles for automated segmentation. In MRI, intensity nonuniformity, motion artifacts, and variability across scanners make boundary delineation inconsistent. CT imaging, while excellent for bone and thoracic regions, suffers from poor soft-tissue contrast and radiation-related restrictions, limiting availability of large annotated datasets. Ultrasound poses difficulties due to speckle noise, shadowing, and operator dependence, which often lead to nonuniform image quality and hinder reproducibility of segmentation models. X-rays, being 2D projections, compress three-dimensional structures, creating overlapping anatomical regions that confuse lesion localization. PET imaging provides functional rather than structural detail; its low spatial resolution and high noise levels complicate precise segmentation. In the case of fundus and retinal photography, variations in illumination, camera specifications, and patient-specific ocular features reduce segmentation stability. Similarly, histopathology images, though extremely detailed, are computationally demanding because of their ultra-high resolution and variation in staining processes.

Beyond modality-specific concerns, several broader challenges persist. Small or diffuse lesions, such as microaneurysms in retinal scans or tumor cores in MRI, are difficult to detect reliably. Severe class imbalance, where pathological regions occupy only a tiny portion of the image, leads models to bias toward healthy structures. Inter-patient variability in anatomy, disease presentation, and demographic factors adds another layer of complexity. Finally, domain shift between datasets collected at different institutions or with different acquisition protocols frequently causes performance drops when algorithms are deployed in real clinical environments.

1.5 Best practices and countermeasures

Rather than searching for one definitive remedy, the field has moved toward integrating diverse techniques to ease the difficulties of segmentation across imaging types. Some of these focus on data preparation, others on improving learning algorithms, and still others on maintaining reproducibility. When employed together, they provide a more balanced and dependable framework for advancing segmentation in practice.

1. *Preprocessing tailored to modality*
   - MRI/CT: intensity harmonization, bias-field correction, and denoising help reduce scanner-related artifacts.
   - Ultrasound: speckle reduction and adaptive filtering improve image clarity; standardized protocols lessen operator dependence.
   - Histopathology: patch-based analysis and stain normalization address large file sizes and color variability.
2. *Dealing with limited data and imbalance*
   - Augmentation strategies (rotation, scaling, elastic deformation, contrast adjustments) expand training variability.
   - Transfer learning leverages pretrained models to reduce the need for massive labeled datasets.
   - Semi-supervised and weakly supervised learning exploit unlabeled or partially labeled samples to ease annotation burdens.
3. *Model design choices*
   - Encoder–decoder frameworks such as U-Net remain the foundation due to their ability to retain both fine and global structures.
   - Attention modules highlight disease-relevant regions, while ensembles improve robustness by combining diverse predictors.
   - Cross-validation and multi-center evaluation ensure generalization across different datasets and institutions.
4. *Benchmarking and reproducibility*
   - Shared datasets (e. g., BRATS, ISIC, REFUGE) and organized challenges provide fair platforms for performance comparison.
   - Public benchmarks encourage transparency, reproducibility, and eventual translation of methods into clinical settings.

1.6 Future directions

The trajectory of medical image segmentation suggests that progress will depend on closing the gap between laboratory models and clinical application. Upcoming work is not limited to precision gains; equal attention is being placed on efficiency, generalization across diverse cases, and reliability in sensitive decision-making contexts. New paradigms in machine learning, coupled with advances in system design and healthcare integration, are likely to define this next phase.

1. *Data efficiency and annotation reduction*
   - Self-supervised and unsupervised learning approaches can help models learn useful features without relying heavily on labeled datasets.
   - Active learning strategies may reduce annotation effort by prioritizing the most informative samples for expert review.

2. *Generalization across domains*
   - Domain adaptation and transfer learning methods will be crucial for handling variations in scanners, acquisition protocols, and populations.
   - Multi-center collaborations can provide diverse data, improving robustness to domain shifts.
3. *Lightweight and real time models*
   - Development of compact architectures suitable for deployment on low-resource clinical systems and mobile devices.
   - Edge computing may allow near real-time segmentation for time-sensitive applications such as surgery or emergency care.
4. *Explainability and trust*
   - Integration of explainable AI techniques can make segmentation outputs more interpretable for clinicians.
   - Visual attention maps and uncertainty estimation will support decision-making and foster clinical acceptance.
5. *Multimodal and holistic analysis*
   - Combining structural (MRI, CT) and functional (PET, ultrasound) imaging can lead to richer, more accurate segmentation.
   - Future models may also integrate imaging with nonimaging clinical data to enhance diagnostic value.
6. *Ethics, fairness, and standardization*
   - Building frameworks that address bias, ensure equitable performance across demographic groups, and adhere to ethical guidelines.
   - Development of global standards for medical image segmentation to support regulatory approval and clinical deployment.

2 Recent advancements in state-of-the-art brain-tumor detection and segmentation

Abstract: Accurate brain-tumor segmentation from MRI is essential for diagnosis and treatment planning. This chapter reviews the evolution of segmentation strategies, from traditional thresholding and clustering to classical machine learning and the current dominance of deep learning. Recent progress with U-Net derivatives, adversarial frameworks, and CNN–Transformer hybrids is highlighted, together with their reported results on benchmark datasets such as BraTS and TCIA. Comparative insights are drawn between 2D and 3D designs, CNNs and Transformers, and single versus ensemble models. Publicly available datasets are summarized, along with their benefits and inherent limitations. Major challenges remain, including heterogeneous tumor appearance, domain variability, and annotation scarcity. Looking ahead, promising directions include self-supervised learning, federated training across institutions, lighter models for clinical use, and explainable AI. Collectively, these developments mark important progress but also emphasize the need for clinically robust and generalizable solutions.

Keywords: Medical image segmentation, Brain tumor segmentation, Biomedical image analysis, Deep learning, MRI, U-Net, Transformers

2.1 Introduction

Brain tumors represent one of the most critical health challenges, with high mortality rates and significant implications for patient quality of life. Early and accurate identification of tumors, along with precise delineation of their subregions, is vital for clinical decision-making in surgery, radiotherapy, and treatment planning [211]. Manual segmentation of brain tumors performed by radiologists is both time-consuming and prone to interobserver variability, creating a strong need for automated, reliable solutions.

Medical imaging plays a central role in the diagnosis and monitoring of brain tumors. Among the available modalities, Magnetic Resonance Imaging (MRI) is the most widely adopted due to its ability to capture high-resolution structural information across multiple contrasts such as T1, T2, FLAIR, and contrast-enhanced T1 (T1c) [311]. Computed Tomography (CT) remains useful in emergency scenarios or when MRI is contraindicated, while Positron Emission Tomography (PET) provides valuable functional and metabolic insights that complement structural imaging. The multimodal nature of brain imaging introduces opportunities for robust tumor characterization, but also poses challenges in terms of data integration and analysis.

In recent years, the rapid development of artificial intelligence (AI) and machine learning (ML) techniques has transformed medical image analysis. Traditional approaches that relied on handcrafted features and statistical models are increasingly

https://doi.org/10.1515/9783111389059-002

being replaced by deep learning (DL) architectures capable of learning hierarchical feature representations directly from imaging data. For brain-tumor segmentation, this shift has resulted in significant performance improvements, particularly with the adoption of convolutional neural networks (CNNs), U-Net variants, generative adversarial networks (GANs), and, more recently, vision transformers and hybrid architectures.

The aim of this chapter is to provide a focused review of recent advancements in brain-tumor detection and segmentation. Emphasis is placed on the evolution of segmentation techniques, the role of publicly available datasets in benchmarking progress, and the analysis of state-of-the-art methods. The chapter concludes with a discussion of ongoing challenges and outlines promising future research directions for this rapidly advancing field.

2.2 Brain-tumor segmentation techniques

2.2.1 Traditional approaches

Before the dominance of machine learning and deep learning, brain-tumor segmentation was mainly attempted using rule-based or classical image processing techniques. These included simple intensity-based thresholding, region growing algorithms, clustering strategies, deformable models, and atlas registration. Thresholding was often used as the starting point, exploiting pixel intensity differences to mark tumor regions. Although conceptually simple, even small variations in image contrast or scanner settings could cause large errors, making this approach fragile. Region growing improved the situation by adding spatial continuity, yet in practice it frequently "leaked" into surrounding structures whenever tumor edges were diffuse or poorly defined. Clustering approaches such as k-means and fuzzy c-means introduced unsupervised grouping of pixels into homogeneous regions. While they sometimes provided a better partition of brain tissues, they often struggled when the tumor shared overlapping intensity values with normal tissues like edema. Deformable contour models (e. g., snakes, level sets) were also popular because of their ability to outline irregular shapes. Their drawback was heavy dependence on initialization and parameter tuning, which made them unstable for routine use. Another line of research used atlas- or template-based registration, where a healthy brain atlas was aligned with the patient scan to detect abnormalities. The concept was appealing, but in reality failed often, since tumor growth deforms brain anatomy significantly, and the registration becomes unreliable. In summary, these early techniques laid the groundwork for automated tumor analysis and demonstrated that automation was possible. However, their lack of robustness, sensitivity to imaging noise, and inability to generalize across patients limited their clinical utility. These shortcomings eventually pushed the field toward learning-based methods.

Sharif et al. [252] integrated fuzzy median filtering, unsupervised clustering, and Gabor texture features within an ELM framework. The approach was computationally efficient but showed reduced reliability when segmenting very small or noisy tumor regions. Arumugam et al. [18] introduced a neuro-fuzzy classifier enhanced through Binary Cuckoo Search optimization. This combination improved segmentation accuracy and stability, although the design required considerable parameter tuning and remained relatively complex. Barzegar and Jamzad [32] designed a weighted label-fusion method that merged atlas priors with patch-based learning. Their framework produced accurate delineations on BRATS datasets, but heavy computation and sensitivity to parameters limited its broader applicability. Shahvaran et al. [250] applied a morphological active contour model guided by k-means and MRF spatial priors. The method converged quickly and adhered well to boundaries, though its dependence on initialization reduced generalizability. Alpar et al. [10] combined Ricker-type wavelet transforms with fuzzy clustering to address low-contrast lesion segmentation. While effective in those settings, validation was restricted, and the method required careful parameter calibration. Kumar and Boda [140] fused active contour models with fuzzy CNN classifiers and employed swarm optimization to tune parameters. This hybrid system improved segmentation accuracy but was computationally demanding and initialization-sensitive. Pruthi et al. [216] incorporated river formation dynamics into active contour evolution, generating smoother tumor boundaries and resilience to noise. However, performance was mainly demonstrated on simulated BrainWeb data rather than large clinical cohorts. Wang et al. [315] proposed a sparse Bayesian decision strategy with joint label fusion, yielding high Dice and sensitivity across tumor regions. The trade-off was weaker boundary precision and substantial computational overhead. Mamatha et al. [169] developed a graph-theory-based segmentation using minimum spanning tree cuts. Their approach captured structural complexity effectively, but external validation was limited and generalization across diverse datasets was not established. Table 2.1 provides an overview of representative traditional segmentation methods, highlighting how each contributed to early progress, while also revealing common drawbacks such as noise sensitivity and poor generalization.

Conventional image processing techniques such as thresholding, clustering, and deformable models provided the first attempts at automated brain-tumor segmentation. While they were straightforward and in some cases effective for capturing broad tumor regions, their performance was highly sensitive to noise, parameter choices, and initialization. Most of these approaches lacked robustness when applied to diverse clinical datasets and were unable to generalize across variations in tumor appearance or imaging protocols. These shortcomings highlighted the need for data-driven frameworks, leading to the adoption of machine learning methods that could exploit extracted features to improve accuracy and adaptability.

Table 2.1: Summary of a few relevant traditional methods for brain-tumor segmentation.

Ref.	Methodology	Advantages	Limitations	Dataset	DSC	Sens.	PPV
[252]	Fuzzy median filter + clustering + Gabor features with ELM/RELM classifiers	Fast training; robust to variations	Struggles with noisy scans; weak in small tumor subregion detection	BRATS 2015	0.89	0.89	0.87
[18]	Neuro-fuzzy classifier with Binary Cuckoo Search	High accuracy; robust optimization with BCS	Complex system design	Kaggle dataset	–	0.88	–
[32]	Semi-supervised Weighted Label Fusion Segmentation using probabilistic graph-based label propagation	Accurate delineation of whole tumor and core regions; combines atlas priors with learning-based fusion	Parameter-sensitive; higher computational demand; limited validation beyond BRATS datasets	BRATS 2015	0.91	0.89	–
[250]	Morphological active contour with k-means and MRF	Fast convergence; robust boundaries	Initialization dependent; limited generalization	BRATS 2013	0.91	0.90	–
[10]	Ricker-type wavelet imaging with Gaussian preprocessing	Effective for low-contrast lesion detection	Limited validation on public datasets	BRATS 2012	0.87	–	–
[140]	Fusion of Active Contour and FCM with J–Tunicate Swarm	Optimization improves accuracy	Higher compute; limited external validation	Kaggle dataset	–	1.0	0.97
[130]	FCM-based level set with kernel mapping	Nonlinear mapping enhances segmentation	Sensitive to initialization; unstable contours	BRATS-2017	0.97	0.98	–
[216]	River Formation Dynamics with Active Contour Model	Smooth and continuous contours; robust to noise	Tested mainly on simulated data	BrainWeb	0.97	0.98	–
[315]	Sparse Bayesian decision framework and joint intensity–spatial label fusion	High Dice and sensitivity	Computationally demanding	BRATS 2017	0.95	0.93	–
[169]	Graph theory–based segmentation with minimum spanning tree graph cuts	Captures complex structures; efficient in handling noisy MRI	Computationally intensive; tested on limited datasets; generalization not validated	BRATS 2018	–	–	–

2.2.2 Machine learning-based approaches

The step beyond basic image processing was the introduction of machine learning (ML). Instead of segmenting tumors purely with rules, ML approaches attempted to learn patterns from annotated scans. The usual workflow was to extract descriptive features from MRI images and then feed them into a classifier. Typical features included: (i) intensity values and their statistical summaries, (ii) texture measures such as gray-level co-occurrence matrices (GLCM) or wavelet coefficients, (iii) shape- or edge-based descriptors designed by domain experts. Various classifiers were then explored. Support Vector Machines (SVM) were widely used, mainly because they handled high-dimensional features relatively well. Random Forests were also popular since their ensemble design could reduce the effect of noisy inputs. Some groups tried simpler tools like k-Nearest Neighbors (k-NN) or logistic regression, while others experimented with hybrid models that combined more than one classifier. These systems did outperform thresholding and clustering, but they carried serious limitations. The outcome often depended on how good the handcrafted features were, and designing those features required considerable expertise. Tumor appearance also varied drastically across patients, so features that worked in one dataset sometimes failed on another. Scalability was another problem: Applying the same pipeline to large, multi-institutional datasets usually revealed its lack of robustness. Overall, ML marked an improvement over rule-based methods, but its dependence on feature engineering kept it from being a long-term solution. This gap prepared the ground for deep learning, where feature extraction and classification could be merged into a single trainable framework.

Rao and Karunakara [226] developed a KSVM–SSD pipeline that combined nonnegative matrix factorization preprocessing, statistical texture descriptors, and kernel SVM classification. Their model achieved strong accuracy and grade-level classification on BRATS data, but the multistage design was parameter-sensitive and complex. Alqazzaz et al. [11] combined deep features from a SegNet encoder with GLCM texture descriptors, followed by decision-tree classification. The approach reduced computational overhead and improved whole-tumor accuracy, though core and enhancing tumor regions were segmented less reliably. Santhosh Kumar and Karibasappa [103] presented a hybrid deep-belief network optimized with an Adaptive Search and Rescue algorithm, integrating wavelet and Gabor features. This boosted segmentation robustness and noise handling, but at the expense of significant computational cost. Sathies Kumar et al. [142] proposed an optimized deep belief network trained with GS-MVO feature selection, incorporating trilateral filtering and fuzzy centroid-based region growing. Their method demonstrated stable convergence and high accuracy on Kaggle MRIs, though performance on challenging tumor subregions was less convincing. Mallick et al. [168] introduced a deep wavelet autoencoder coupled with a deep neural network for tumor classification. By compressing feature dimensions, the model improved accuracy over conventional DNNs, but was tested only on a small clinical dataset.

Soltaninejad et al. [267] applied supervoxel partitioning with Gabor and statistical features, using Random Forests for classification. The design efficiently captured multimodal information and reduced computational cost, though segmentation of small tumor cores remained a weakness. Kumar et al. [141] proposed the IBRDM framework, fusing radiomic descriptors with DWT features and employing ensemble classifiers such as RF, DT, and SVM. Their pipeline produced high accuracy with strong feature optimization but required heavy computation and was limited to BRATS validation. Dai et al. [62] introduced a gradient-guided active-learning framework where a VAE-driven-data manifold suggested informative samples for annotation, later segmented with U-Net. The model reduced annotation cost drastically, though it depended heavily on expert involvement and the quality of the learned manifold. Alqhtani et al. [12] combined CLAHE-based preprocessing, FCM segmentation, and SVM classification to achieve high sensitivity and specificity on a CE-MRI dataset. While efficient and accurate across multiple tumor types, its generalizability beyond the tested cohort remains unproven. Kollem et al. [134] proposed a pipeline combining fuzzy C-means segmentation with an SVM classifier whose parameters were tuned by an opposition-based grey-wolf optimizer. The framework effectively captured multiscale tumor structures and yielded robust tissue classification, although the design remained relatively complex and parameter-dependent. Asiri et al. [19] developed a two-stage computerized approach, where image enhancement relied on adaptive Wiener filtering, neural networks, and ICA, followed by SVM-based segmentation and classification. Their method achieved strong performance across meningioma and pituitary tumor cases on MRI data, but its clinical generalization beyond the tested cohorts is yet to be demonstrated. Table 2.2 outlines key machine-learning frameworks for brain-tumor segmentation, showing how feature-based classifiers improved accuracy over conventional image processing, yet continued to face challenges of complexity, feature dependency, and dataset sensitivity.

While machine learning offered a step forward compared to conventional image processing, many of its methods depended on manually designed features and showed inconsistent results when tumor size, shape, or contrast varied. In several cases, the models performed well only on the datasets they were trained on, and the overall pipelines became intricate, requiring extensive tuning. Scalability to high-dimensional multimodal scans remained limited, and robustness across different clinical settings was difficult to achieve. These issues gradually opened the way for deep learning, where both feature discovery and classification are learned jointly within a single architecture.

2.2.3 Deep learning-based approaches

The real breakthrough in brain-tumor segmentation came with deep learning (DL). Unlike earlier ML systems that depended on manually designed features, DL models learn feature hierarchies directly from the data itself. This ability has made them the pre-

Table 2.2: Summary of a few significant and relevant machine learning-based methods for brain tumor segmentation.

Ref.	Methodology	Advantages	Limitations	Dataset	DSC	Sens.	PPV
[226]	KSVM-SSD framework with NMF preprocessing	Effective multistage classification	Complex pipeline; sensitive to parameter tuning	BRATS 2018	0.99	1.00	0.99
[11]	Hybrid method combining SegNet-based learned features with GLCM	Reduces computational load via ROI masking	Lower accuracy for core and enhancing regions	BRATS 2017	0.98	–	–
[103]	Hybrid DBN with Adaptive Search and Rescue	Improved accuracy; robust noise handling	Complex pipeline; higher computational cost	BRATS 2015	0.98	0.99	0.96
[142]	GS-MVO optimized Deep-Belief Network with skull stripping	Robust feature optimization with GS-MVO	Limited validation on challenging subregions (edema, necrosis)	Kaggle dataset (270 images)	0.95	0.94	0.95
[168]	Deep Wavelet Autoencoder for feature compression combined with DNN	Reduces feature dimensionality	Computationally intensive	TCIA (19 Patients)	0.93	0.92	–
[267]	Supervoxel segmentation with statistical features and Random Forests	Efficient feature representation via supervoxels	Lower Dice for very small cores; supervoxels near boundaries may misclassify	BRATS 2013 (30 cases)	0.89	0.96	0.99
[141]	Radiomics + DWT-based fusion of multimodal MRI	Strong feature optimization; robust cross-validation	Complex, computationally heavy pipeline	BRATS 2018	0.99	–	–
[62]	Gradient-guided active learning framework	Reduces annotation cost significantly	Relies on quality of VAE-learned manifold	BRATS 2019	0.77	–	–
[12]	CLAHE + diffusion filtering, Fuzzy C-Means + SVM	High sensitivity/specificity; fast processing (0.42s)	Parameter-sensitive	CE-MRI (233 patients)	0.96	0.97	–
[134]	FCM with SVM optimized by Opposition-based GWO Optimization	Captures multiscale tumor details; robust tissue classification	Complex pipeline	BRATS 2021			
[19]	Adaptive Wiener filtering, with SVM	Applied to multiple tumor types	Complex model architecture	MRI Nanfang Hospital China	0.98	0.99	–

ferred choice for almost all recent work in the field. A few broad categories of architectures are discussed as follows.

2.2.3.1 Convolutional neural networks (CNNs)

Early studies used 2D CNNs that operated slice-by-slice or on small patches extracted from MRI scans. While effective, this approach sometimes lost the 3D context. Later, researchers moved to 3D CNNs, which capture volumetric information and provide better delineation of irregular tumor regions. Standard backbones such as ResNet, DenseNet, and VGG have also been adapted for segmentation tasks. One consistent drawback, however, is the heavy demand on memory and computation, especially for 3D volumes.

Deng et al. [68] integrated a heterogeneous CNN with a CRF–RRNN to enhance spatial context modeling in MRI tumor segmentation. This design improved boundary precision across multiple modalities but required extensive computational resources and was sensitive to patch selection. Myronenko and Hatamizadeh [188] advanced a 3D encoder–decoder CNN with ResNet blocks and hybrid loss functions, achieving top performance in the BraTS 2019 challenge. While highly accurate, the model's complexity and GPU demands limited its clinical scalability. Zhou et al. [350] introduced OM-Net, a one-pass multitask network with cross-task guided attention to refine coarse-to-fine tumor segmentation. The model reduced system complexity compared to cascaded approaches but still depended on large-scale training and incurred heavy computational costs. Huang et al. [108] developed GammaNet, embedding adaptive gamma correction with Dense-ASPP modules to achieve intensity invariance. The approach improved robustness to MRI intensity variation and tumor-detail capture, though external validation remained limited and training costs were high. Liew et al. [151] proposed CASPIANET++, a channel–spatial asymmetric attention network combined with Noisy Student curriculum learning for semi-supervised training. It delivered high Dice scores and reduced parameter counts but remained sensitive to training schedules and lacked widespread validation. Rai et al. [224] presented a hybrid UnetResNext-50, which fused a U-Net backbone with a ResNext-50 encoder. The system attained very high accuracy and Dice scores, outperforming standard U-Net variants, but evaluation was restricted to a single dataset without broader clinical testing. Sasank and Venkateswarlu [242] integrated a tumor growth-prediction model with a full-resolution CNN, leveraging texture features and metaheuristic optimization. Their method preserved spatial detail and achieved strong accuracy across BraTS datasets but was computationally demanding and highly parameter-sensitive. Zhang et al. [347] introduced MSMANet, a multiscale mesh aggregation network with residual inception modules and attention-based deep supervision. The architecture improved multi-scale feature fusion and robustness to noise, though resource intensity and limited 3D scalability posed challenges. Sun and Wang [280] proposed ASCNN, an application-specific CNN with reduced layers and Full-ReLU activation for efficiency. Their design achieved competitive Dice scores with far fewer parameters, but validation was confined to BraTS and its generalization across diverse cohorts remains uncertain.

Recently, Kuang et al. [138] proposed a lightweight hybrid model (LW-CTrans) combining CNN and Transformer modules was introduced with dynamic stems, MPConv blocks, and MVPFormer pooling to capture both local and global features. It achieved competitive Dice scores across three public datasets and proved effective for small lesion segmentation, though external clinical validation is still required. Balamurugan and Gnanamanoharan [27] designed a hybrid deep CNN with LuNet classifier, integrating LOG-based preprocessing, FCM–GMM segmentation, and VGG-16 feature extraction. Their method reached very high accuracy and reduced training time, but testing was confined to limited MRI datasets, raising concerns about scalability. Valenkova et al. [306] implemented an ensemble strategy unifying SegResNet, UNETR, and SwinUNETR through fuzzy rank-based nonlinear aggregation achieved a Dice of 0.885 on BraTS 2023. The approach demonstrated robustness and statistical significance but incurred high computational costs and lacked evaluation on diverse external datasets.

Table 2.3 summarizes recent CNN-driven frameworks for brain-tumor segmentation, showing how convolutional encoder–decoder models, residual networks, and attention-augmented CNNs have substantially advanced segmentation performance, while still facing challenges such as high computational demands, limited interpretability, and restricted generalization across diverse datasets.

2.2.3.2 U-Net and its variants

The U-Net architecture has become the reference model in medical segmentation. Its encoder–decoder design, with skip connections, enables that both global structure and fine details can be preserved. Over time, many variants have appeared:
- 3D U-Net for volumetric data,
- Attention U-Net to emphasize tumor regions,
- Nested U-Net (U-Net++) for better feature fusion,
- Residual U-Net combining ResNet blocks with U-Net's design.

These extensions dominate benchmarks like BRATS dataset, though they still face difficulties in handling very small tumor subregions.

Zhang et al. [341] proposed AGResU-Net, which integrated residual blocks and attention gates within the U-Net architecture. This design improved delineation of smaller tumors and produced higher Dice scores than baseline models, though it was computationally intensive and confined mainly to 2D slice analysis. Huang et al. [109] advanced this line with GCAUNet, a residual U-Net incorporating multiscale inputs and group cross-channel attention. The framework was effective at recovering tumor boundary details and boosting segmentation accuracy, but its parameter count was high and slice-based training limited full 3D context. Jiang et al. [123] introduced DDU-Net, a

Table 2.3: Summary of a few significant and relevant deep learning methods based on Convolutional Neural Network (CNN) for brain-tumor segmentation.

Ref.	Methodology	Advantages	Limitations	Dataset	DSC	Sens.	PPV
[68]	Hybrid model with Heterogeneous CNN and CRF-RRNN	Captures spatial context and improves boundary accuracy	Computationally intensive; sensitive to kernel size	BRATS 2013	–	–	–
[188]	3D encoder–decoder CNN with ResNet blocks	Robust across subregions	High computational demand; complex architecture	BRATS 2019	0.88	–	–
[350]	One-pass Multi-task Network (OM-Net) with Cross-task Guided Attention	Addresses class imbalance; reduces model complexity compared to cascades	Computational cost significant	BRATS 2015	0.87	0.88	0.89
[108]	Adaptive Gamma Correction for intensity invariance	Improves generalizability across MRI intensities	High computational cost	TCGA-LGG (110 patients)	0.85	0.87	–
[151]	CNN with Channel–Spatial Asymmetric Attention layers	Improves generalization with limited labeled data	Complex attention modules	BRATS 2018	0.91	0.92	–
[224]	Hybrid CNN + U-Net with ResNext-50 encoder	Robust feature extraction	Tested only on a single dataset	TCGA-LGG (110 patients)	0.95	0.89	0.90
[242]	Full Resolution CNN integrated with tumour growth	Preserves spatial resolution	Computationally heavy	BRATS 2018	0.95	0.87	–
[347]	Multiscale mesh aggregation with Res-Inception	Improves multiscale feature representation	Extension to 3D networks limited by resource needs	BRATS 2018	0.89	0.89	–
[280]	Application-Specific CNN with 7 custom layers	Computational efficient	Performance sensitive to training quality	BRATS 2018	0.89	0.88	–
[27]	Hybrid DCNN with enhanced LuNet classifier	Reduced training time due to lightweight LuNet design	Performance may degrade with large cohorts	MRI datasets	0.98	0.99	0.99
[138]	Lightweight CNN– Transformer with dynamic stem	Robust across three medical imaging tasks	Still limited by dataset coverage	Brain tumor dataset	0.83	–	–
[306]	Ensemble of SegResNet, UNETR, and SwinUNETR	Robust to data disturbances	Computationally intensive due to multiple CNN	BRATS 2023	0.88	–	–

dual-stream decoding U-Net with one branch dedicated to edge features and another to semantic segmentation. This approach enhanced boundary precision and handled class imbalance better, but the two-branch structure increased training complexity and evaluation was limited to BraTS datasets. Wang et al. [319] developed CLCU-Net, which connected multiple levels of U-Net with a selective attention module and deep supervision. The model achieved superior Dice and precision scores while maintaining efficient inference, though it still relied on 2D slices and required more memory for attention modules. Cınar et al. [57] combined DenseNet121 as an encoder with a U-Net decoder, producing a hybrid DenseNet121-U-Net. Their approach alleviated vanishing gradient problems and achieved high Dice across tumor regions, but preprocessing was intensive and the evaluation was confined to BraTS 2019. Liu et al. [159] presented S2MetricUNet, a U-Net variant enhanced with scale-adaptive voxel-metric loss and super-voxel features. This design improved segmentation of tumor core and enhancing regions, while reducing computational cost compared to MetricUNet, yet precision was slightly lower and validation was limited to BRATS 2019.

Moreover, Bhatti et al. [39] introduced FF-UNet, which integrates feature fusion with CNN components and dropout layers to enhance segmentation accuracy. Their approach consistently achieved above 98 % accuracy with high precision and Jaccard scores, although computational demands and lack of large-scale clinical testing remain limitations. Kundu et al. [144] proposed FFLUNet, a lightweight U-Net variant that uses multi-view feature fusion and adaptive weighting in skip connections. With only 1.45M parameters and a nearly fivefold faster inference speed compared to nnUNet, the model is suitable for real-time use, though its Dice scores were not explicitly reported and external validation is pending. Pani and Chawla [207] developed a hybrid approach that combines transfer learning, self-supervised learning, and a 3D U-Net with attention modules. The framework reduced reliance on annotated data and achieved a Dice of about 0.90, though the training pipeline is intricate, and evaluation was mainly restricted to BraTS datasets. Lin and Chen (2024) [152] proposed MM-UNet, which leverages module- and scale-based cross-attention to fuse global and local information. Tested on BraTS 2020, the model outperformed several state-of-the-art baselines with a Dice score of around 0.85, but required substantial computational resources and remained dataset-specific. Aboussaleh et al. [4] introduced Inception-UNet, an enhanced U-Net model using Inception blocks in skip connections to improve multi-scale feature representation. It achieved Dice scores of 0.87, 0.85, and 0.83 on BraTS 2020, 2018, and 2017 respectively, though its more complex skip design increases computational overhead and external validation was not reported.

Table 2.4 presents representative U-Net variants applied to brain tumor segmentation. These models extend the original architecture with residual connections, attention mechanisms, and dense or dual-decoder designs, demonstrating improved accuracy and boundary preservation. Despite their strong performance on BRATS datasets, they remain challenged by computational demands and limited external validation.

Table 2.4: Summary of a few significant and relevant deep learning methods based on UNet for brain tumor segmentation.

Ref.	Methodology	Advantages	Limitations	Dataset	DSC	Sens.	PPV
[341]	U-Net with residual modules and attention gate	Improved segmentation of small-scale tumors	Computationally demanding	BRATS 2018	0.876	0.88	0.85
[109]	Residual U-Net with detail recovering path	Effectively preserves tumor boundary details	Slice-based training may miss full 3D spatial context	BRATS 2018	0.90	0.88	–
[123]	Dual-stream decoding U-Net combining semantic and edge extraction branches	Enhances boundary precision by learning explicit edge features	Computational overhead due to dual-branch design	BRATS 2017	0.904	0.911	–
[319]	Cross-level connected UNet with segmented attention	Effectively leverages multi-scale feature connections	Losing inter-slice context	BRATS 2018	0.88	0.85	0.91
[57]	DenseNet121 + UNet with patch-based preprocessing	Alleviates vanishing gradient	Requires extensive preprocessing	BRATS 2019	0.95	0.95	–
[159]	UNet with scale-adaptive voxel-metric loss	Improves Dice on tumor core and enhancing tumor	Still computationally heavy compared to plain UNet	BRATS 2019	0.83	0.89	0.81
[144]	Lightweight U-Net variant with multi-view feature fusion block	Suitable for real-time resource-constrained deployment	External/clinical testing needed	Brain tumor MRI	–	–	–
[207]	Hybrid 3-D UNet using transfer learning and self-supervised learning	Reduces dependency on large annotated datasets	Training pipeline more complex	3D MRI datasets	0.90	–	–
[152]	Multi- scale and multi- module cross attention UNet	Effectively fuses multi-scale features	Computationally intensive due to cross attention layers	BRATS 2020 (369 cases)	0.85	0.88	0.85
[4]	Improved U-Det with Inception blocks	Effective for multi-scale tumor segmentation	Higher computational cost than standard U-Net	BRATS 2017	0.83	–	–

2.2.3.3 GAN-based models

Generative Adversarial Networks (GANs) have been used to refine segmentation outputs. In this setup, a generator predicts the tumor mask, while a discriminator evaluates how realistic it looks. This adversarial process can sharpen boundaries and reduce false positives. GANs have also been used for data augmentation by producing synthetic MRI scans. On the downside, GAN training is unstable and often requires careful tuning.

Qin et al. [219] designed a dual adversarial domain-adaptation framework using two student U-Nets guided by consistency constraints. This approach successfully reduced domain shift across tumor grades and imaging modalities, achieving higher Dice scores on cross-subtype segmentation, although the multi-loss setup increased training complexity. Zeng et al. [339] proposed 3V3D, a three-view adversarial model with dense skip connections and PatchGAN supervision to exploit contextual differences across MRI slices. The method improved boundary recognition and robustness on BraTS and ADNI1 datasets, but training was computationally expensive and generalization beyond these cohorts was not explored. Wu et al. [325] introduced SD-GAN, which modeled bilateral brain symmetry using a conditional GAN to detect tumors as deviations from normal anatomy. This unsupervised approach reduced reliance on annotated data and achieved competitive results with supervised baselines, though it struggled with midline tumors that appear symmetrical. Neelima et al. [194] proposed Optimal DeepMRSeg combined with a GAN classifier optimized using a hybrid Sailfish Political Optimizer (SPO) and CAViaR-SPO strategy. Their framework achieved accuracy above 91 % on BRATS2018 and Figshare datasets, but the pipeline was computationally heavy and overly complex for clinical deployment. Sun et al. [279] combined perceptual and dual-scale cycle-consistent GANs with a 3D ResU-Net to handle multimodal tumor and stroke segmentation. The model effectively synthesized missing modalities and preserved lesion boundaries, yet its multistage design was resource-intensive and evaluated mainly on BraTS2015 and ISLES2015-SISS.

Table 2.5 summarizes adversarial learning approaches for brain-tumor segmentation. GAN-based frameworks enhance robustness by refining boundaries, addressing domain shifts, and synthesizing missing modalities. While these models often outperform CNN baselines, they typically involve multistage training and higher computational overhead, limiting broader clinical use.

2.2.3.4 Transformers and hybrid approaches

Recently, vision transformers have been explored for their strength in modeling long-range dependencies. Pure transformer models are rare in segmentation, but hybrid CNN–Transformer networks, such as TransUNet or Swin-UNet, have achieved strong results in brain tumor studies. They combine CNNs' ability to capture local detail with transformers' global attention. These methods are promising but computationally expensive and relatively new in clinical research.

Ghazouani et al. [86] combined Swin Transformer blocks with enhanced local self-attention (ELSA) and CNN modules in an encoder–decoder setup. Their design improved both long-range dependency modeling and local feature capture, producing an average Dice of nearly 90 % on BraTS 2021, though it was more computationally demanding than CNN baselines. Pan et al. [204] proposed VcaNet, a hybrid 3D CNN–Transformer framework integrating enhanced convolutional (ENCO) modules, a multiscale Transformer bottleneck (MSCTrans), and CBAM attention in the decoder. The model successfully fused local volumetric features with global dependencies and outperformed CNN and Transformer baselines on BraTS datasets, but its layered architecture required substantial resources. Hatamizadeh et al. [97] developed Swin UNETR, a hierarchical Transformer encoder with shifted-window self-attention linked to a CNN decoder through skip connections. This method achieved top-tier results in the BraTS 2021 challenge, with Dice scores above 0.92 for whole-tumor segmentation, but at the cost of high parameter counts and computational intensity. Liu et al. [160] introduced TransSea, a hybrid network with semantic awareness that used mutual attention, semantic guidance, and semantic integration modules to bridge CNN and Transformer features. This approach significantly improved segmentation on BraTS 2020 and 2021, yet the inclusion of multiple semantic modules increased design complexity and training cost. Zeng et al. [340] proposed DBTrans, a dual-branch Vision Transformer designed for multimodal brain tumor segmentation. The network combines a local branch using shifted-window self-attention with a global branch employing shuffle-window cross-attention, enabling it to capture both fine details and global context more efficiently. An additional cross-attention branch in the decoder and channel-attention modules further enriched feature integration. Experiments showed that DBTrans surpassed existing CNN and Transformer baselines, though its dual-branch design increased architectural complexity and computational requirements.

Overall, Transformer and hybrid CNN–Transformer models demonstrate strong potential by capturing both global context and local detail, addressing weaknesses of pure CNNs or ViTs. While they consistently outperform traditional architectures on BraTS datasets, most require large computational budgets and remain limited in external validation, suggesting a gap between research performance and practical clinical deployment. Table 2.6 provides an overview of Transformer-driven and hybrid CNN–Transformer models. By capturing long-range dependencies while retaining local detail, these architectures achieve state-of-the-art accuracy on BraTS datasets. However, their complexity, large parameter counts, and reliance on high-end computational resources remain major barriers to real-world adoption.

Table 2.5: Summary of a few significant and relevant deep learning methods based on Generative Adversarial Network (GAN) for brain tumor segmentation.

Ref.	Methodology	Advantages	Limitations	Dataset	DSC	Sens.	PPV
[219]	Dual adversarial domain-adaptation with two U-Nets	Mitigates domain shift across tumor grades	Computationally expensive	BRATS 2019	0.88	–	–
[339]	Three-view contextual cross-slice difference network	Improves edge features and robustness	High computational demand	BRATS 2019	0.87	–	–
[325]	Symmetry-driven Generative Adversarial Network	Competitive with supervised U-Net	Fails with midline tumors due to symmetry	BRATS 2018	0.61	0.61	0.73
[194]	Sailfish Political Optimizer + CAViaR-SPO-based GAN	Enhances classification using hybrid GAN training	Complex multi-stage pipeline	BRATS 2018	–	0.92	–
[279]	GAN with perceptual cycle and dual-scale consistency	Generates realistic missing modalities	Higher computational needs	BRATS 2015	0.690	0.68	0.69

Table 2.6: Summary of a few significant and relevant deep learning methods based on transformers and hybrid approach for brain tumor segmentation.

Ref.	Methodology	Advantages	Limitations	Dataset	DSC	Sens.	PPV
[86]	Hybrid Swin Transformer–CNN with Enhanced Local Self-Attention blocks	Captures long-range dependencies with Swin attention	More computationally intensive than CNNs	BRATS 2021 (1251 cases)	0.89	–	–
[204]	Hybrid 3D CNN–Vision Transformer with enhanced convolution	Captures both local volumetric and global contextual features	Architecture is complex	BraTS 2020	–	–	–
[97]	Hierarchical Swin Transformer encoder with shifted window self-attention	Captures long-range dependencies and multiscale features	High parameter count (62M) and computational cost	BraTS 2021	0.92	–	–
[160]	Hybrid CNN–Transformer encoder–decoder with Semantic Mutual Attention	Effectively integrates global Transformer features with local CNN features	Architecture complexity with multiple semantic modules	BraTS 2020	–	–	–
[340]	Dual-branch Vision Transformer with cross-attention and channel attention	Balances local and global feature extraction with reduced computational cost	Complex dual-branch design	BRATS	–	–	–

2.2.3.5 Strengths and limitations

Deep learning approaches clearly outperform earlier methods in accuracy, robustness, and adaptability to multimodal data. They have become the de facto standard for brain-tumor segmentation challenges. At the same time, a few concerns remain:
- They require large annotated datasets,
- GPU and memory requirements are high,
- Performance often drops when applied to data from different hospitals or scanners (domain shift),
- Interpretability is limited, which affects clinical trust.

Despite these challenges, DL-based methods represent the most successful step so far, and their evolution continues rapidly with newer architectures and learning strategies.

2.3 Publicly available datasets

The development of advanced brain-tumor segmentation methods has been strongly supported by the availability of large-scale, curated datasets. Public datasets enable fair benchmarking, reproducibility of results, and foster innovation through community challenges. Among them, the BRATS challenge datasets have emerged as the de facto standard for brain-tumor segmentation research, complemented by other repositories such as TCIA, BrainWeb, and smaller online collections.

2.3.1 BraTS challenge datasets

The Brain-Tumor Segmentation (BraTS) [177] challenges, running annually since 2012, provide multi-institutional MRI scans across four modalities: T1, T1c, T2, and FLAIR. Ground-truth labels typically include whole tumor, tumor core, and enhancing tumor regions. These datasets have standardized preprocessing pipelines (e. g., skull stripping, co-registration, isotropic resampling) and evaluation metrics such as Dice score and Hausdorff distance, ensuring fair comparison across methods. However, the curated nature and preprocessing may introduce domain bias, limiting direct generalization to raw clinical settings.

2.3.2 TCIA collections

The Cancer Imaging Archive (TCIA) [243] hosts several brain tumor-imaging collections, such as TCGA-GBM and TCGA-LGG cohorts, which include multimodal MRI, clinical, and sometimes genomic data. These datasets are invaluable for radiogenomic studies, though segmentation labels are not consistently available, making them less straightforward for benchmarking segmentation methods.

2.3.3 BrainWeb

BrainWeb [59] provides simulated MRI datasets with precise ground truth labels and configurable noise/artifact levels. It is widely used for algorithm validation under controlled conditions. The main limitation is that simulated images lack the variability of real-world data, so models trained exclusively on BrainWeb may not generalize well to clinical cases.

2.3.4 Other benchmark repositories

Several smaller datasets are available through platforms such as Figshare and Kaggle, offering annotated brain tumor MRI scans. While often useful for prototyping or educational purposes, these datasets tend to have limited scale and annotation quality. ISLES datasets [166], although focused on ischemic stroke, are occasionally used for testing cross-domain generalization of segmentation models.

2.3.5 General observations and limitations

Public datasets have been pivotal in driving progress, yet challenges remain. Most datasets originate from research settings with standardized acquisition, which does not reflect the heterogeneity of real-world clinical imaging. Rare tumor subtypes are underrepresented, and variations in annotation protocols across datasets complicate cross-comparison. Addressing these gaps is essential for translating segmentation models into reliable clinical tools.

2.4 Discussion

Examining the spectrum of existing segmentation methods makes it clear that the field has advanced quickly, yet a number of obstacles remain. Researchers continue to debate the balance between accuracy, efficiency, and practicality, with each family of models offering distinct benefits and trade-offs.

2.4.1 Comparative insights

One of the most common design decisions concerns whether to rely on 2D or 3D processing. Two-dimensional models can be trained with modest hardware and are faster at inference, but they overlook volumetric continuity that is often crucial in delineating tumor boundaries. By contrast, 3D networks provide richer contextual information and usually yield better structural consistency, though they require more memory and training time.

Convolutional networks have served as the backbone of most segmentation pipelines because of their ability to capture local structural details. Still, their limited receptive field constrains global reasoning. Transformers, and especially CNN–Transformer hybrids, attempt to address this by integrating local and long-range dependencies in a single framework. These designs often report higher benchmark scores, but they bring increased parameter counts and training cost. Ensemble models push accuracy further by merging predictions from multiple systems, but such strategies are rarely practical outside research settings due to their heavy computational footprint.

2.4.2 Key challenges

Beyond architecture choice, the task itself poses difficulties. Tumor regions are highly heterogeneous, and their boundaries are often indistinct, especially in the case of small enhancing components. Patient-to-patient and scan-to-scan variation further complicates generalization since models tuned to one dataset frequently degrade on another. Reliable labels remain another bottleneck since volumetric annotation is labor-intensive and requires domain expertise. Finally, domain-shift across institutions and scanner types remains a stubborn problem, limiting direct translation of research models into clinical environments.

Overall, while new methods have undoubtedly raised segmentation performance, the field must move beyond benchmark accuracy. Robustness to domain variability, reduced reliance on annotations, and computationally efficient designs are likely to be the deciding factors in whether these algorithms can achieve meaningful impact in everyday clinical workflows.

2.5 Conclusion and future directions

This chapter reviewed the evolution of brain tumor-segmentation methods, tracing the shift from traditional image processing and machine learning techniques toward deep learning approaches that now dominate the field. Early clustering- and atlas-based strategies provided useful insights but lacked robustness, while classical machine learning improved performance through handcrafted features. The arrival of deep learning, particularly convolutional and U-Net-based architectures, has dramatically advanced segmentation accuracy, making them the backbone of state-of-the-art systems. More recently, hybrid networks combining CNNs, Transformers, and attention mechanisms have emerged, showing promise in modeling both local details and long-range dependencies.

Several clear trends are evident. U-Net and its numerous variants remain the most widely adopted backbone, often extended with residual connections, attention modules, or dense skip pathways. Hybrid CNN–Transformer models and multimodal fusion frameworks represent the next stage of progress, aiming to balance contextual reasoning with computational efficiency. Ensemble approaches continue to raise benchmark scores, though at the cost of resource demands.

Looking forward, a number of research directions are likely to shape the field. Self-supervised and weakly supervised methods could alleviate the reliance on expert-labeled datasets, lowering annotation costs. Federated learning offers a pathway for privacy-preserving training across institutions, helping to address domain shift and data scarcity. Lightweight architectures capable of real-time inference are essential for integration into clinical workflows, especially in resource-limited environments. Finally, explainable AI will play a key role in improving interpretability and building trust among clinicians, ensuring that automated systems support rather than obscure medical decision-making.

Overall, while substantial progress has been achieved, translating these advances into reliable clinical practice will require addressing both technical and practical challenges, with future research expected to focus on making models more generalizable, efficient, and transparent.

3 Advanced deep learning architectures for automated brain-tumor segmentation

Abstract: This chapter explores the application of advanced deep learning architectures for fully automated brain-tumor segmentation in magnetic resonance imaging. The discussion follows the progression from early convolutional designs to adaptive pipelines and more recent Transformer-driven frameworks, with attention also given to adversarial learning approaches. Reported results from the literature are synthesized to illustrate how these models have been evaluated within the BRATS challenges, emphasizing differences in methodology, dataset usage, and evaluation protocols. Beyond architectural details, the chapter highlights recurring experimental practices, such as preprocessing strategies, training configurations, and the choice of evaluation metrics. In doing so, the review offers a consolidated perspective on current segmentation techniques, while also identifying open challenges related to computational efficiency, robustness, and clinical applicability. The chapter concludes by outlining potential directions for future research, including data-efficient learning, lightweight model design, and integration of multimodal information for more reliable clinical translation.

Keywords: Medical image segmentation, Computer-aided diagnosis, Brain tumor segmentation, Deep learning architectures, Magnetic resonance imaging (MRI)

3.1 Introduction

The previous chapter provided a comprehensive overview of state-of-the-art brain-tumor segmentation methods, highlighting how architectures such as U-Net variants, attention-based networks, GANs, and Transformer-driven designs have shaped recent progress in the field. While such reviews are essential for understanding theoretical advances, researchers and practitioners often face challenges when translating these ideas into practical workflows. For many, the real difficulty lies not in grasping the high-level concepts, but in understanding how to implement, train, and evaluate these models under realistic conditions.

This chapter shifts the focus from a theoretical survey to practical demonstration. Rather than introducing new algorithms, it aims to illustrate how advanced deep learning frameworks can be applied to brain-tumor segmentation in a reproducible manner. Using representative architectures, including 3D U-Net [358], Attention U-Net [198], nnU-Net [116], Swin UNETR [97], TransBTS [318], and GAN-based models [195], we provide a step-by-step view of their implementation and show example outcomes on widely used datasets such as BRATS [177]. The intention is not to propose novel benchmarks, but to highlight how these models behave in practice, what trade-offs they introduce, and how their performance can be analyzed for educational and clinical insight.

https://doi.org/10.1515/9783111389059-003

By presenting results as illustrative case studies, this chapter serves as a bridge between a literature review and real-world experimentation. Readers are guided through the essential elements of training, evaluation, and comparison, enabling them to appreciate both the strengths and the limitations of current methods. In this way, the chapter complements the review in Chap. 2 by offering a practice-oriented perspective, preparing the ground for specialized directions such as transfer learning, which will be discussed in the subsequent chapter.

3.2 Overview of selected deep learning-based methods

To demonstrate the application of advanced deep learning in brain-tumor segmentation, this chapter focuses on a small but representative set of models. These architectures cover three major categories: convolutional encoder–decoders, self-configuring pipelines, and hybrid networks that combine convolutional layers with Transformers or adversarial learning. Each is briefly introduced, with schematic diagrams provided for clarity.

3.2.1 3D U-Net

The 3D U-Net [358] is an extension of the original U-Net that replaces 2D operations with volumetric convolutions. This modification allows the model to capture spatial continuity across MRI slices, making it particularly suitable for 3D medical images. Its symmetric encoder–decoder structure with skip connections ensures that both local details and global context are preserved.

The 3D U-Net extends the well-known 2D U-Net to volumetric inputs, enabling it to process MRI scans as complete 3D structures. Its encoder–decoder design enables hierarchical feature extraction, while maintaining spatial continuity across slices. In practice, it has become the benchmark against which new approaches are compared, particularly in the BRATS challenges. However, the need for large memory resources often forces researchers to rely on patch-based training, which can fragment contextual information in very large scans. Fig. 3.1 shows the architecture of 3D U-Net encoder-decoder with downsampling, upsampling, and skip connections for volumetric medical image segmentation.

3.2.2 Attention U-Net

The Attention U-Net [198] introduces attention gates within the skip connections. Originally proposed for pancreas segmentation, the Attention U-Net has since been adapted for brain-tumor tasks. These gates learn to highlight regions of interest-such as tumor

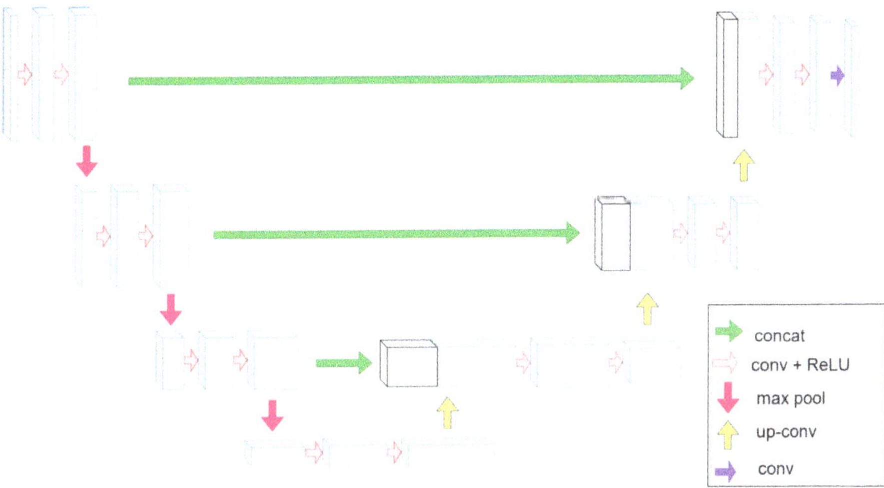

Figure 3.1: Overview of the 3D U-Net architecture. The model consists of an encoder that extracts hierarchical volumetric features through convolution and pooling and a decoder that progressively upsamples and refines the feature maps. Skip connections between matching encoder and decoder stages help preserve spatial details. This schematic highlights the overall workflow; numerical details such as kernel sizes and feature dimensions are omitted for simplicity.

boundaries–while suppressing irrelevant background information. The mechanism improves the model's focus without adding substantial computational cost, making it useful for segmenting structures with irregular shapes.

Attention U-Net integrates attention gates into skip connections so that the model can automatically highlight informative tumor regions, while suppressing irrelevant tissue. This modification improves delineation in low-contrast or highly irregular tumors without fundamentally altering the U-Net backbone. The benefit comes at the cost of extra computations, although the overhead remains modest compared to the performance gain in complex segmentation tasks.

3.2.3 nnU-Net

Unlike fixed architectures, nnU-Net [116] is a self-configuring framework that automatically adapts its preprocessing, network depth, and training parameters to the dataset. It has become a de facto baseline in medical image-segmentation challenges, largely due to its adaptability and robust performance across multiple tasks. For this chapter, we use its default configuration as generated for the BRATS dataset.

The nnU-Net framework distinguishes itself by automatically adapting preprocessing, network depth, and training configurations to the dataset in use. This self-configuring property has led to its dominance in several segmentation benchmarks where manual tuning is impractical. Yet, such automation reduces transparency for researchers who wish to explicitly control hyperparameters, and the end-to-end pipeline may require longer training times on limited hardware.

3.2.4 Swin UNETR

Swin UNETR [97] integrates Swin Transformer blocks into an encoder–decoder pipeline. While convolutional layers capture local textures, the Transformer blocks model long-range dependencies, enabling the network to better handle tumors of varying sizes and shapes. Its hierarchical design also supports multiscale representation learning. The Swin UNETR architecture combines patch embedding with hierarchical Transformer encoding and U-Net style decoding supported by skip connections as shown in Fig. 3.2.

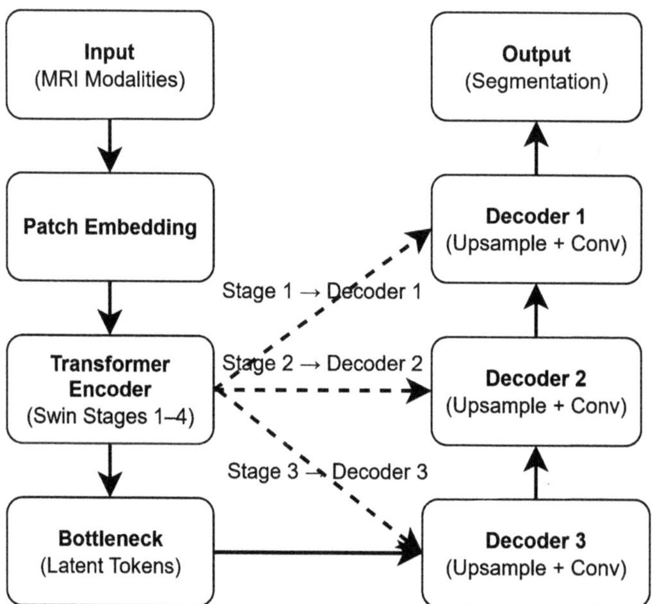

Figure 3.2: Schematic representation of the Swin UNETR framework. Input MRI volumes are partitioned into patches and embedded as tokens, which are processed through hierarchical Swin Transformer encoder stages (Stages 1–4). The bottleneck latent tokens are progressively decoded by U-Net style decoder blocks with upsampling and convolution layers. Dashed arrows denote skip connections from encoder Stages 1–3 to the corresponding decoders, enabling multiscale feature fusion. Low-level details are omitted for clarity. Adapted from Hatamizadeh et al. [97].

Swin UNETR employs shifted-window self-attention to combine global context modeling with hierarchical feature extraction. The model integrates these features into a U-Net style decoder, which makes it well-suited for tumors spanning diverse shapes and sizes. While effective, its reliance on transformer blocks implies high memory consumption and a strong dependency on large training datasets to ensure stable convergence, making it less accessible for smaller clinical cohorts.

3.2.5 TransBTS

The TransBTS model [318] combines a 3D CNN encoder with Transformer modules that operate on flattened feature maps. This design enables the model to balance local detail preservation with global contextual reasoning. The CNN backbone extracts spatial features, while the Transformer captures dependencies across slices, making it a flexible hybrid for volumetric segmentation. Figure 3.3 presents the simplified version of the TransBTS model that integrates a CNN encoder with a transformer bottleneck and a CNN-based decoder.

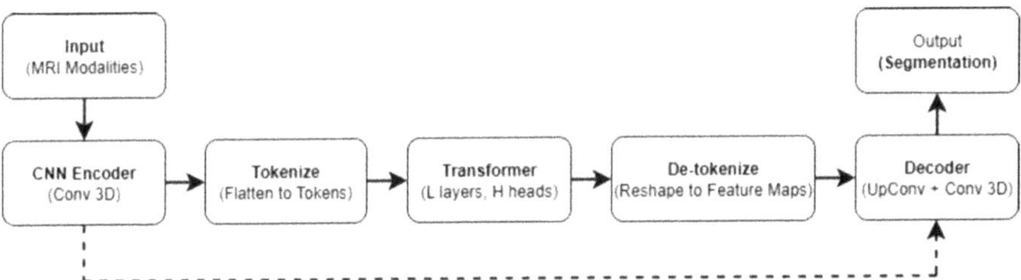

Figure 3.3: Schematic representation of the TransBTS framework. A 3D CNN encoder extracts volumetric features which are flattened into tokens and processed by a Transformer bottleneck. The outputs are reshaped into feature maps and passed through a CNN-based decoder to produce the segmentation. In the full model, the decoder incorporates multiple upsampling stages with U-Net style skip connections from the encoder; these are omitted here for clarity and readability. Adapted from Wang et al. [318].

TransBTS couples a 3D CNN encoder with a transformer bottleneck, merging local spatial details with long-range dependency modeling. This balance between convolutional and Transformer-based processing makes it possible for it to better capture both fine and contextual structures within MRI volumes. However, the added complexity of coordinating CNN and Transformer modules demands careful hyperparameter tuning, which may limit its straightforward adoption.

3.2.6 GAN-based segmentation

Generative adversarial networks introduce a discriminator that evaluates how realistic the predicted segmentation masks are. By training a generator (segmentation network) and a discriminator in tandem, the method enforces structural plausibility and sharper boundaries. Although more complex to train, GAN-based approaches can reduce false positives and improve segmentation quality [107, 329]. Nie et al. [195] proposed a GAN-based segmentation setup introducing adversarial feedback in addition to supervised learning, as shown in Fig. 3.4.

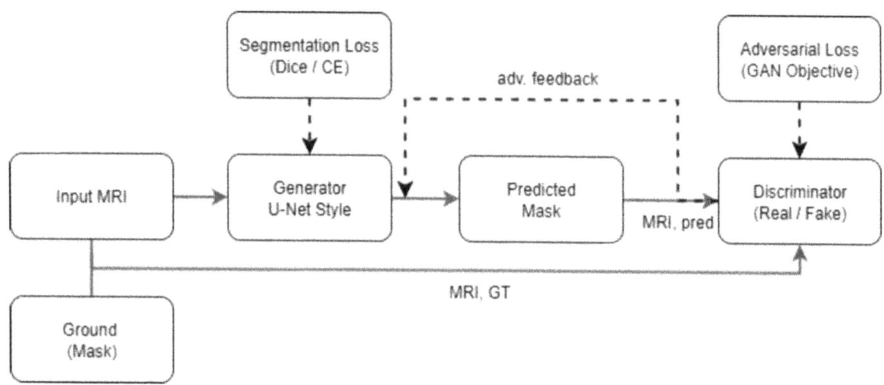

Figure 3.4: Conceptual diagram of a GAN-based segmentation framework. The generator (U-Net style encoder–decoder) predicts a segmentation mask from the input MRI. The discriminator receives pairs formed by concatenating the MRI with either the predicted mask or the ground-truth mask, and learns to distinguish real from generated. The generator is optimized with both supervised segmentation loss (e. g., Dice or cross-entropy) and adversarial feedback from the discriminator. In inference, only the generator is used. Adapted from Nie et al. [195].

GAN-based approaches treat segmentation as an adversarial game between a generator and a discriminator. The generator produces segmentation masks, while the discriminator evaluates their realism against ground-truth masks paired with MRI scans. This setup encourages sharper and more plausible outputs, especially around tumor boundaries. Still, adversarial training is known for instability, and results may vary substantially depending on the relative weighting of supervised and adversarial loss terms.

It is worth noting that some of the GAN-based methods referenced here, such as the context-aware framework by Nie et al. [195] and the SegAN architecture proposed by Xue et al. [329], were originally introduced for broader medical image synthesis and segmentation tasks rather than specifically for brain-tumor analysis on the BRATS dataset. These studies are included in this discussion because they represent influential milestones in bringing adversarial learning into medical imaging, and they have subsequently inspired adaptations to brain tumor segmentation. Their role in this chapter is therefore conceptual and methodological, highlighting how GANs were first explored in the field, while later tumor-focused studies (e. g., Huang et al. [107]) demonstrate their direct application to BRaTS.

In summary, these models illustrate the variety of strategies used in current deep learning research, from convolutional baselines to attention mechanisms, automated

Table 3.1: Comparison of selected deep learning models for brain-tumor segmentation.

Model	Type	Strengths	Limitations
3D U-Net	Encoder–decoder (CNN)	Strong volumetric baseline; robust feature hierarchy	High memory demand; often requires patching
Attention U-Net	CNN with attention gates	Highlights relevant tumor regions; suppresses noise	Slight increase in computation; marginal complexity
nnU-Net	Self-configuring CNN framework	Adapts automatically to datasets; consistent across benchmarks	Less user control; longer training on modest hardware
Swin UNETR	Transformer encoder + CNN decoder	Captures long-range dependencies; good multi-scale fusion	Heavy GPU requirements; data-hungry for stability
TransBTS	Hybrid CNN + Transformer	Balances local detail and global context	Complex to optimize; sensitive to hyperparameters
GAN-based Segmentation	Generator–discriminator	Produces realistic, sharper masks; reduces false positives	Training instability; loss balancing critical

pipelines, and hybrid networks. Their inclusion here provides a balanced view of how different design choices can affect tumor-segmentation performance in practice. The extended descriptions and the comparative summary in Table 3.1 make it evident that no single architecture universally dominates in brain-tumor segmentation. Each framework introduces unique advantages, such as volumetric continuity in 3D U-Net, selective feature enhancement in Attention U-Net, adaptability in nnU-Net, long-range dependency modeling in Swin UNETR, or hybrid integration in TransBTS. Likewise, adversarial learning offers sharper boundaries through GAN-based designs. However, these strengths are balanced by specific drawbacks ranging from heavy memory demands to instability in training. To assess how these architectural characteristics translate into practical performance, the following section presents the experimental setup and results obtained on standard benchmark datasets.

3.3 Comparative analysis of experimental setups

Most of the experimental evaluations reported in the literature are anchored in the BraTS challenges, which remain the most widely adopted benchmark for brain-tumor segmentation [24, 25, 177]. These datasets provide four MRI modalities (T1, T1c, T2, and FLAIR), together with expert labels for the main tumor compartments: whole tumor (WT), tumor core (TC), and enhancing tumor (ET). Because of their standard format and yearly updates, BraTS datasets (2018–2021) have become the reference point for comparing segmentation methods. A smaller set of publications supplemented BraTS with in-house clinical scans to demonstrate that the proposed models could generalize beyond a controlled challenge setting.

Despite differences in model design, the data preparation steps across these studies are remarkably consistent. Scans are typically resampled to achieve uniform voxel resolution, intensities are normalized—often using z-score statistics within the brain mask—and volumetric patches are extracted to circumvent GPU memory limits. Patch sizes around 128^3 voxels are commonly reported for 3D models such as U-Net derivatives, Swin UNETR, or TransBTS. Additional preprocessing techniques, including bias-field correction, intensity clipping, or even modality dropout, appear in selected studies. To counter the risk of overfitting inherent to limited medical datasets, authors frequently rely on heavy data augmentation, combining random flips and rotations with elastic warping or intensity perturbations.

Training choices vary with the architecture. For example, the original 3D U-Net was trained with stochastic gradient descent and momentum, while more recent CNN and hybrid approaches such as TransBTS generally use the Adam optimizer, with learning rates typically in the 10^{-3} to 10^{-4} range. Transformer-based designs, such as Swin UNETR, favor AdamW together with learning-rate scheduling (e. g., cosine decay) and mixed-precision training on large-scale GPUs. Loss formulations are usually a mixture of Dice and cross-entropy to address class imbalance and boundary accuracy. Adversarial models add a GAN loss component, which demands careful balancing between generator and discriminator updates. A different philosophy is seen in nnU-Net, which configures hyperparameters automatically, employing strategies like five-fold cross-validation to stabilize training.

Evaluation procedures are also broadly aligned. The Dice similarity coefficient (DSC) [69] remains the dominant metric, reported separately for WT, TC, and ET. Most studies additionally include the 95th percentile Hausdorff distance (HD95) [110] to evaluate boundary quality. Sensitivity and specificity are occasionally added to assess clinical reliability in terms of false positives and false negatives. Recent works, such as Swin UNETR, also emphasize efficiency by reporting model size and inference speed.

In summary, although individual optimizers, patch sizes, or learning schedules may differ, these experimental setups converge around a common backbone defined by the BraTS benchmarks. This shared ground allows for meaningful cross-study comparisons, which will be examined in the subsequent section.

3.4 Results and discussion

Table 3.2 collects outcomes reported for several segmentation frameworks various different BraTS editions. The listed Dice and Hausdorff metrics provide a sense of how accuracy has improved over successive years, though direct comparison is limited by differences in dataset versions and evaluation protocols.

Table 3.2: Performance of representative models on the BraTS datasets [24, 25, 177] Values correspond to Dice similarity coefficient (DSC) [69] for the three tumor subregions; Hausdorff distance (HD95) [110] is shown where available. Results are taken from the original publications and may not be directly comparable across years due to differences in dataset versions and evaluation protocols.

Model	Dataset	WT	TC	ET	HD95 (mm)
3D U-Net [358]	BraTS 2017	0.85	0.75	0.70	–
nnU-Net [116]	BraTS 2021	0.92	0.81	0.88	5
TransBTS [318]	BraTS 2021	0.91	0.89	0.86	4
Swin UNETR [97]	BraTS 2021	0.93	0.91	0.89	3
GAN-based Seg. [107]	BraTS 2020	0.90	0.81	0.70	4

Taken as a whole, the results reflect a steady refinement of methodology. Early deep learning models confirmed that volumetric MRI data could be segmented effectively using convolutional architectures. Later approaches placed greater emphasis on automated configuration and systematic training, which reduced dependence on manual design choices. More recent developments have introduced mechanisms for capturing long-range dependencies and structural consistency, leading to higher reported performance on multiple tumor subregions. Alongside these, alternative strategies based on adversarial training have explored various ways of enforcing anatomical plausibility.

A consistent message across challenge reports is that numerical scores alone do not fully capture model behavior. Factors such as stability across patients, sensitivity to small regions, and the overall coherence of predicted tumor shapes are repeatedly discussed in the literature. These aspects underline the importance of complementing benchmark accuracy with evidence of robustness and clinical reliability [131].

3.5 Conclusion and future directions

This chapter reviewed representative deep learning approaches for automated brain tumor segmentation and summarized their reported performance on the BraTS benchmarks. The comparison indicates a gradual evolution in methodology, moving from early volumetric convolutional networks to adaptive frameworks such as nnU-Net and, more recently, to transformer-based models that incorporate long-range contextual reasoning. While each family of methods contributed incremental improvements, the overall trajectory demonstrates a clear rise in segmentation accuracy and boundary quality over successive years. Nevertheless, the analysis also highlighted that accuracy metrics alone provide an incomplete picture, as practical deployment depends equally on robustness, interpretability, and stability across diverse imaging conditions.

Looking forward, several areas are likely to shape the next stage of research. First, the computational demands of Transformer-driven models remain a barrier for routine use, creating a need for lighter architectures and more efficient training strategies. Second, data scarcity continues to limit progress, making semi-supervised, self-supervised, and federated learning approaches particularly attractive for leveraging unlabeled or distributed clinical datasets. Third, integrating multimodal information—such as genomic markers, clinical variables, or longitudinal scans—could provide richer context and more clinically meaningful outputs. Finally, questions of reproducibility, fairness, and transparency will become increasingly important as these systems move closer to clinical practice.

In summary, while deep learning has already achieved substantial progress in brain-tumor segmentation, ongoing research is expected to balance two parallel objectives: further improvements in benchmark accuracy and the development of models that are computationally efficient, generalizable across sites, and aligned with clinical decision-making needs. Meeting both goals will be essential for translating technical advances into tangible benefits for patients and healthcare providers.

The discussion in this chapter has centered on fully automated segmentation frameworks developed and benchmarked on the BraTS challenges. While these models demonstrate strong performance, they typically require large labeled datasets and extensive training resources. In many practical situations, however, annotated data are scarce and computational budgets are limited. These constraints have motivated increasing interest in transfer learning, where knowledge from pretrained models or related tasks is adapted to brain-tumor analysis. The following chapter explores this direction in detail, presenting transfer learning strategies for both tumor detection and segmentation, and examining how they can reduce data requirements while maintaining competitive accuracy.

4 Transfer learning for brain-tumor detection and segmentation

Abstract: Work on brain-tumor detection and segmentation often struggles with the same bottleneck: limited and inconsistent MRI data. To address this, many groups now rely on transfer learning, where networks trained for one problem are repurposed for another. In this area, some approaches keep it simple, using CNN backbones as feature extractors, while others employ U-Net encoders or more recent transformer blocks. Hybrid combinations are also common. In this chapter, we bring together these studies, pointing out how weights are reused, what datasets have been tested, and how the results compare. We also comment on challenges, such as domain differences between pretraining and medical data, and suggest likely directions that could improve reliability in the future.

Keywords: Transfer learning, Brain tumor segmentation, MRI, Deep learning, Pretrained models

4.1 Introduction

Medical image analysis has become one of the most active research areas in artificial intelligence (AI), particularly with the increasing use of deep learning for diagnosis and treatment planning. In the context of brain-tumor detection and segmentation, magnetic resonance imaging (MRI) is the modality of choice due to its ability to capture detailed soft-tissue information. However, deep learning models typically require very large and diverse datasets to achieve robust performance. Unlike natural image repositories, medical imaging datasets are usually limited in size, suffer from class imbalance, and may differ in acquisition protocols across institutions. These challenges often restrict the ability to train deep networks from scratch and can lead to overfitting and poor generalization [26, 177].

Transfer learning has emerged as a powerful strategy to overcome these limitations by reusing knowledge learned from large-scale pretrained models. Instead of starting from random initialization, models trained on well-established datasets, such as ImageNet, are adapted to medical tasks with relatively small modifications. This approach not only accelerates convergence but also helps the network capture low-level and high-level features relevant for medical image analysis [205, 321]. In medical imaging, several studies have confirmed the benefits of fine-tuning pretrained networks compared to training from scratch, showing improved efficiency and accuracy even with limited datasets [259, 286].

In this chapter, we specifically focus on the role of transfer learning for brain-tumor detection and segmentation using MRI. Numerous families of deep learning models—

https://doi.org/10.1515/9783111389059-004

including convolutional neural networks (CNNs), U-Net and its variants, Transformer-based architectures, and hybrid ensembles—have been successfully adapted through transfer learning. A systematic review of these approaches highlights not only their effectiveness but also their evolution over time. This section sets the stage for discussing how these models have been designed, fine-tuned, and evaluated for clinical applications.

4.2 Background on transfer learning

Transfer learning refers to the process of reusing knowledge gained in one task or domain and applying it to another, often related, task. Instead of training a model from scratch, which requires large annotated datasets and significant computational resources, transfer learning enables researchers to leverage pretrained models and adapt them to target applications with relatively fewer data samples. This strategy has gained strong relevance in medical imaging, where dataset scarcity, annotation cost, and institutional variability are persistent challenges [205, 321].

Two common strategies are widely adopted: feature extraction and fine-tuning. In feature extraction, a pretrained network (e. g., ResNet, VGG, DenseNet) is used as a fixed feature encoder, and only the final classification or segmentation layers are retrained on the target medical dataset. Fine-tuning, by contrast, allows some or all of the pretrained layers to be updated during training, thus adapting the learned features to the specific characteristics of medical images. The choice between these approaches depends on dataset size, task similarity to the pretraining domain, and available computational resources [259, 286].

A crucial enabler of transfer learning is the availability of large-scale benchmark datasets. The most influential is ImageNet, containing over 14 million annotated natural images across 1,000 categories [67]. Although ImageNet represents a domain different from medical imaging, its pretrained weights capture low-level structures such as edges, textures, and shapes that generalize well across tasks. These features serve as effective initialization points for medical imaging, including brain-tumor detection and segmentation.

Over the past decade, several backbone architectures trained on ImageNet have become standard for transfer learning:

1. *AlexNet (2012)*: First deep CNN breakthrough on ImageNet, demonstrating that multilayered convolutional architectures can outperform traditional methods by a large margin [137].
2. *VGGNet (2014)*: Introduced very deep (16–19 layer) architectures with uniform small convolutional kernels, popular for transfer learning due to its straightforward design [264].

3. *GoogLeNet/Inception (2015)*: Proposed inception modules to efficiently capture multiscale features within the same layer, reducing computation while preserving accuracy [284].
4. *ResNet (2015)*: Introduced residual connections, enabling very deep networks (50+ layers) to train effectively and becoming one of the most widely used backbones for transfer learning [98].
5. *DenseNet (2017)*: Connected each layer to all subsequent layers, promoting feature reuse and improving gradient flow, which proved beneficial in medical imaging tasks [106].
6. *EfficientNet (2019)*: Introduced compound scaling to balance accuracy and efficiency, offering strong performance with fewer parameters compared to earlier CNNs [288].

As seen in Table 4.1, each backbone introduced a distinct architectural idea—ranging from residual connections in ResNet to compound scaling in EfficientNet—that directly influenced their adoption in transfer learning pipelines for brain-tumor detection and segmentation. Their pretrained weights, often initialized on ImageNet, provide strong

Table 4.1: Major backbone models used for transfer learning in medical imaging, highlighting their key contributions, relevance to transfer learning, and architectural innovations.

Model	Year	Contribution & Role in Transfer Learning	Architectural Highlights
AlexNet [137]	2012	First deep CNN to succeed on ImageNet; introduced ReLU and dropout; inspired early medical TL studies using pretrained CNNs for feature extraction.	8 layers (5 conv + 3 FC); overlapping max-pooling; dropout regularization.
VGGNet [264]	2014	Demonstrated that very deep networks with uniform convolutional kernels generalize well; widely used as encoder backbone in segmentation models.	16–19 layers; uniform 3×3 conv; 2×2 pooling.
GoogLeNet / Inception [284]	2015	Efficient multiscale feature learning with fewer parameters; used in TL tasks for classification and medical imaging benchmarks.	Inception modules with parallel 1×1, 3×3, 5×5 filters; factorized conv layers.
ResNet [98]	2015	Enabled training of very deep models via residual connections; most popular backbone for TL in medical imaging due to stability and accuracy.	Skip connections; 50–152 layers; solved vanishing gradient issue.
DenseNet [106]	2017	Improved feature reuse and gradient flow; adopted in medical TL for efficient representation learning with fewer parameters.	Dense connectivity between layers; compact parameterization.
EfficientNet [288]	2019	Balanced accuracy and efficiency via compound scaling; increasingly used in TL for medical imaging where computational cost is a concern.	Jointly scales depth, width, and resolution for optimal efficiency.

starting points for downstream tasks such as brain-tumor detection and segmentation. These architectures, originally designed for natural image tasks, have been repeatedly applied in medical imaging studies, including brain-tumor MRI classification, detection, and segmentation. More recent advances extend beyond CNNs to Vision Transformers (ViT) pretrained on large-scale datasets, but CNN backbones remain foundational in transfer-learning pipelines. In the specific case of brain-tumor research, transfer learning provides substantial benefits because clinical MRI data are heterogeneous, limited in quantity, and expensive to annotate. By leveraging pretrained models, researchers can accelerate convergence, improve generalization, and achieve performance gains even when working with relatively small or imbalanced datasets [26, 177].

The backbone models outlined in Table 4.1 provide the foundation for most transfer learning strategies applied in medical image analysis. In brain-tumor detection and segmentation, these architectures are either directly fine-tuned or serve as encoders within more complex designs, such as U-Net variants or hybrid frameworks. Building on this foundation, the following sections categorize transfer-learning approaches into four main groups: CNN-based, U-Net–based, Transformer-based, and hybrid methods highlighting how each family has been adapted to MRI data and evaluated in the context of brain-tumor analysis.

4.3 Representative approaches in brain-tumor imaging

4.3.1 CNN-based transfer learning

Convolutional neural networks (CNNs) have been widely adopted as backbone architectures for brain-tumor analysis, with transfer learning enabling adaptation from large-scale natural image datasets to medical imaging tasks. Several studies demonstrate that pretrained CNNs such as ResNet, VGG, and Inception can provide strong feature extraction capabilities for classification and detection. A comparative overview of CNN-based transfer-learning approaches applied to brain-tumor MRI analysis is presented in Table 4.2, summarizing the key methodologies, datasets, and reported performance metrics.

Sharma et al. [254] applied a modified ResNet50 framework where the final layer was replaced by additional dense layers, and transfer learning was used to adapt pretrained weights to MRI classification. Data augmentation and dropout improved generalization, enabling sensitivity and precision values up to 1.0 on the Kaggle brain MRI dataset. However, the model's applicability is constrained by its reliance on a single dataset and absence of external clinical validation. Liu et al. [157] introduced Transition Net, a hybrid framework that employs a 2D Swin Transformer backbone with a 3D Transition Decoder for multimodal MRI tumor segmentation. While the approach stabilizes segmentation across subregions, its effectiveness remains dependent on backbone selection and it inherits computational demands typical of Transformer–CNN hybrids.

Table 4.2: Summary of representative CNN-based transfer-learning approaches for brain-tumor detection and classification using MRI.

Ref.	Methodology	Advantages	Limitations	Dataset	DSC	Sens.	PPV
[254]	Modified ResNet50 with transfer learning	Robust against overfitting	Grade-level classification limited	Kaggle Dataset	–	0.89	0.89
[157]	2D Swin Transformer backbone with a 3D decoder and weighted region loss	Effective handling of ET/TC regions with weighted loss	Domain shift from 2D ImageNet to 3D MRI can reduce transferability	BraTS 2019	0.91	–	–
[334]	CNN + VGG-16 TL + Ensemble	Robust feature extraction via pretrained VGG-16	Small dataset	Custom dataset (253 scans)	–	0.94	0.96
[239]	AlexNet CNN with transfer learning	XAI provided interpretability	Performance gains limited on T1 MRIs	Glioma MRI	–	–	–
[162]	CNN + FCN with RGB fusion	Improving classification and segmentation	Misclassification in meningioma / metastasis	Internal dataset	0.93	0.99	0.98
[283]	VGG19 with block-wise fine-tuning	Avoids handcrafted features	Dataset imbalance affects meningioma class	CE-MRI (233 patients)	–	–	–
[262]	InceptionV3 pretrained on ImageNet	PSO improved convergence and robustness	Tested only on Kaggle dataset	Kaggle (6,328 scans)	–	0.99	0.99
[128]	Transfer learning with ResNet152 and GoogleNet	Optimized ResNet152 + SVM achieved high accuracy	Focus on 2D MRIs only	Custom MR dataset	–	0.96	0.97

Younis et al. [334] combined CNN features with a VGG-16 transfer-learning backbone and ensemble learning to classify brain tumors. While effective, the study's limited dataset and absence of external validation constrain its broader clinical applicability. Rustom et al. [239] introduced a novel transfer-learning strategy by seeding AlexNet with camouflage animal detection features before fine-tuning on brain-tumor MRI classification. The approach significantly boosted accuracy on T2-weighted MRIs (92.2 %) compared to baseline training and was complemented by explainable AI tools such as GradCAM and DeepDream to visualize network decision-making. However, results on T1 MRIs were less improved, and the study remains constrained by dataset imbalance and lack of external validation.

Moreover, Lu et al. [162] introduced a CNN–FCN framework leveraging non-contrast MRI by fusing T1w, T2w, and their average into RGB channels. Models such as ResNet50 and Darknet53 demonstrated strong performance. While promising for patients unsuitable for contrast-enhanced imaging, the method showed challenges with tumor heterogeneity and lacked large-scale clinical validation. Swati et al. [283] developed a VGG19-based content-based image retrieval system for brain tumors, using blockwise fine-tuning and closed-form metric learning. The approach achieved a mean average precision of 96.13 % on CE-MRI data without relying on handcrafted features. While highly effective for retrieval, its scope was limited to 2D slices and retrieval tasks rather than direct segmentation. Siddique et al. [262] introduced a PSO-optimized InceptionV3 framework for brain-tumor classification using MRI. Transfer learning from ImageNet was combined with Particle Swarm Optimization to fine-tune hyperparameters, achieving an outstanding 99.33 % validation accuracy on a Kaggle dataset. While results demonstrate strong classification capability, the reliance on a single public dataset and near-perfect training accuracy raise concerns about overfitting and generalizability. Kaur and Mahajan [128] investigated transfer learning using ResNet152 and GoogleNet as feature extractors, combined with traditional classifiers such as SVM and KNN. The ResNet152–SVM pipeline delivered the best results, achieving 98.53 % accuracy, 96.52 % sensitivity, and an F1 score of 97.4 %. While effective, the study was constrained to a limited dataset and focused on 2D MRI, with no external validation to confirm clinical robustness.

4.3.2 U-Net variants with pretrained encoders

U-Net and its variants remain central to medical image segmentation, and their performance has been further enhanced through transfer learning from pretrained encoders such as VGG, ResNet, and EfficientNet. These approaches leverage pretrained weights to accelerate convergence and improve segmentation accuracy in limited-data scenarios. Table 4.3 provides a consolidated comparison of U-Net–based transfer learning methods applied to brain-tumor MRI segmentation, highlighting their architectural choices, datasets, and reported performance.

Table 4.3: Summary of U-Net–based transfer learning approaches for brain tumor segmentation.

Ref.	Methodology	Advantages	Limitations	Dataset	DSC	Sens.	PPV
[190]	nnU-Net architecture with transfer learning	Reproducible open-source pipeline	Limited availability of pediatric datasets	51 patients	0.71	–	–
[230]	3D U-Net pipeline with optimized pre/post-processing	Clear ablations for preprocessing, loss design, and TL	Sensitivity to thresholding	BRATS 2023	0.79	–	–
[36]	Enhanced ResUNet with EfficientNetB0 encoder	Improved multiscale contextual learning	HD95 inconsistent across subregions	BRATS 2020	0.90	–	–
[213]	VGG19-based U-Net with transfer learning	Robust feature extraction from pretrained VGG19	Lacks multimodal/3D validation	TCGA-LGG (120 patients)	0.96	0.98	0.95

Nalepa et al. [190] employed an nnU-Net–based segmentation framework, transferring knowledge from BraTS glioblastoma cases to pediatric optic pathway gliomas (OPGs). The fine-tuned model achieved Dice scores of 0.78 on the training cohort and 0.71 on the test set, showing that transfer learning can partially overcome small dataset limitations. However, segmentation accuracy remained lower than for adult glioblastoma, highlighting the challenge of limited pediatric data. Ren et al. [230] proposed a 3D U-Net framework enhanced with optimized preprocessing and a compound Dice+Focal+Edge loss, combined with transfer learning from BraTS Challenge 1 to Challenge 2. The approach improved lesion-wise Dice scores across cohorts (0.79 for C1, 0.72 for C2, 0.74 for C3), demonstrating the value of decoder-freezing strategies under limited data. However, performance remains sensitive to post-processing and has yet to be validated on external datasets. Behzadpour et al. [36] proposed an enhanced ResUNet architecture that integrates an EfficientNetB0 encoder, channel attention, and ASPP for brain-tumor segmentation. The model achieved strong performance on BraTS 2020, with Dice scores of 0.903 for WT and 0.851 for TC, outperforming baseline ResUNet variants. While effective, the approach adds computational cost and remains to be validated on external clinical datasets. Pourmahboubi et al. [213] developed a VGG19-based U-Net model with transfer learning and Focal Tversky loss for brain tumor segmentation. On the TCGA-LGG dataset, the method achieved excellent results, including Dice 0.968, Recall 0.982, and Precision 0.954, surpassing standard U-Net and alternative backbones. However, reliance on 2D slices and a single dataset limits its generalizability to broader clinical applications.

4.3.3 Transformer-based transfer learning

Transformers have recently been introduced into brain-tumor analysis, offering advantages in modeling long-range dependencies and global feature representations. When combined with transfer learning, Vision Transformers (ViT), Swin Transformers, and their hybrids have demonstrated competitive or superior performance compared to CNN-based methods. An overview of Transformer-based transfer-learning methods for brain-tumor MRI is summarized in Table 4.4, capturing their architectures, datasets, and key performance trends.

Lai et al. [145] applied the Vision Mamba (Vim), a state-space model adapted for visual tasks, to brain-tumor classification using transfer learning. Compared against ViT, Swin Transformer, EfficientNet, Inception-V3, and ResNet50, the Vim model achieved 100 % test accuracy on a six-class MRI dataset, while remaining lightweight and computationally efficient. Despite these results, the study was limited by dataset size and scope, requiring further validation across broader clinical cohorts. El Joudi et al. [73] applied SegFormer with an adaptive transfer learning strategy for handling imbalanced pixel-level segmentation tasks. The framework achieved Dice scores of 0.84–0.88 across multiple benchmarks, showing its potential to improve class balance without manual

Table 4.4: Summary of Transformer-based transfer-learning approaches for brain-tumor analysis.

Ref.	Methodology	Advantages	Limitations	Dataset	DSC	Sens.	PPV
[145]	Vision Mamba (Vim) with transfer learning	Lightweight and computationally efficient	Limited tumor categories	Kaggle dataset	–	0.99	0.99
[73]	SegFormer-based adaptive transfer learning	Robustness under class imbalance	Focused on synthetic datasets	Synapse, ACDC, BRATS (subset)	0.84	–	–
[202]	Swin Transformer with Hybrid Shifted Windows Multi-Head Self-Attention	Robust feature extraction with ResMLP	Lacks external multicenter evaluation	Kaggle Brain MR	–	0.99	0.99
[8]	Vision Transformer with transfer learning	Captures long-range dependencies	Requires large computational resources	BRATS 2023	–	0.99	0.97
[210]	Hybrid Swin + CSwin Transformer with virtual adversarial training	Robust against intensity variations	High computational cost	FeTS 2022 (1251 volumes)	0.91	0.91	0.91
[228]	Fine-tuned Vision Transformers	High accuracy due to self-attention	Meningioma class still harder to classify accurately	7023 MRI images	–	99.5	99.6
[97]	Hierarchical Swin Transformer encoder + CNN decoder with skip connections	Top-performing method in BRATS 2021 validation	Requires heavy computation	BRATS 2021	0.92	–	–

weighting. Pacal [202] introduced a Swin Transformer framework enhanced with Hybrid Shifted Windows Self-Attention (HSW-MSA) and Residual MLP blocks for brain-tumor classification. Using transfer learning and augmentation on a Kaggle dataset, the model achieved 99.92 % accuracy, outperforming CNN and ViT baselines. Despite its efficiency and strong results, the study is limited to a single dataset without external validation. Ali [8] introduced ViT-BT, a Vision Transformer model enhanced with transfer learning from VGG16 and EfficientNet-B7. The model achieved 98.17 % accuracy on BraTS 2023, showing strong precision and recall by leveraging global self-attention alongside pretrained CNN features. While effective for major tumor types, the framework demands high computational resources and struggles with rare cases, limiting broader clinical generalization

Peiris et al. [210] proposed CR-Swin2-VT, a hybrid volumetric transformer combining Swin and CSwin window attention with virtual adversarial training for brain-tumor segmentation. Evaluated on FeTS 2022, it achieved Dice scores of 0.9138 (WT), 0.8540 (TC), and 0.8171 (ET), demonstrating improved robustness to interinstitutional variability. However, the model's high computational demand and limited clinical validation remain challenges. Reddy et al. [228] introduced fine-tuned Vision Transformer (FTVT) models for multiclass brain-tumor classification, benchmarking them against CNN baselines such as ResNet50 and EfficientNet. The FTVT-l16 variant achieved consistently high precision and recall across tumor types. Despite their strong performance, transformer models required substantial compute and showed relatively weaker accuracy on meningioma cases. Hatamizadeh et al. [97] introduced Swin UNETR, combining a hierarchical Swin Transformer encoder with a CNN-based decoder for multi-modal 3D MRI brain-tumor segmentation. While highly accurate, the approach required multi-GPU training and ensemble strategies, raising concerns about computational efficiency and deployment feasibility.

4.3.4 Hybrid methods

Recent research trends also explore hybrid transfer-learning frameworks that combine the strengths of multiple paradigms, such as CNNs, Transformers, GANs, and optimization-based strategies. These methods aim to address dataset imbalance, improve robustness, and enhance interpretability in clinical contexts. Representative hybrid transfer-learning strategies, combining CNNs, Transformers, GANs, or optimization modules, are compared in Table 4.5, providing insight into their reported performance and applicability.

Napravnik et al. [192] introduced RadiologyNET, a large-scale pseudo-labeled medical dataset used to pretrain foundation models across CNNs and U-Net variants. When applied to downstream tasks including brain-tumor MRI, RadiologyNET-pretrained models achieved performance comparable to ImageNet baselines and showed clear advantages in resource-limited training conditions. However, gains were less pronounced

Table 4.5: Summary of hybrid transfer-learning approaches that integrate CNNs, Transformers, GANs, or optimization frameworks for brain-tumor detection and segmentation.

Ref.	Methodology	Advantages	Limitations	Dataset	DSC	Sens.	PPV
[192]	RadiologyNET foundation pretrained models	Better alignment with domain-specific tasks	Pseudo-labeling dataset may affect generalization.	COVID-19, Brain MRI	0.71	–	0.99
[94]	AlexNet pretrained on ImageNet	Effective in resource-limited scenarios	AUC still moderate compared to the state of the art	BRATS 2019 (335 patients)	–	–	–
[174]	RadImageNet pretrained CNNs	Better interpretability and domain alignment	Trained on a single-clinic dataset with potential bias	RadImageNet	–	–	–
[121]	Hybrid CNN–Transformer model	Robust long- and short-range feature extraction	Some instability on unseen test data	BRATS 2021	0.92	–	–
[337]	Hybrid CNN–Transformer with explainability	Combines CNN local feature extraction with Transformer global context	Higher computational demand	BraTS 2019	0.38	–	–
[290]	EfficientNetV2 and Vision Transformer	Ensemble boosts accuracy	Needs large annotated data	Kaggle Brain MRI	–	0.96	0.96
[102]	Hybrid ensemble (VGG16 + InceptionV3 + VGG19) with transfer learning	Robust multiclass classification	Dataset size relatively small	Custom dataset (3,264 images)	–	0.95	0.95

when sufficient annotated data was available, highlighting both the promise and limitations of domain-specific foundation pretraining. Hao et al. [94] proposed a hybrid framework that integrates transfer learning with active learning to reduce annotation requirements for brain-tumor MRI classification. Using a pretrained AlexNet and uncertainty-based sampling, the method achieved a test AUC of 0.829 on BraTS 2019, while cutting labeling costs by up to 70 %. Mei et al. [174] introduced RadImageNet, a large-scale medical-imaging pretraining dataset of 1.35M radiologic images spanning CT, MRI, and ultrasound. CNNs pretrained on RadImageNet consistently outperformed ImageNet baselines, especially on small datasets, yielding up to 9.4 % AUC improvement and higher Dice for lesion localization. While demonstrating strong domain-specific transferability, the approach's reliance on a single-clinic dataset may limit generalization. Jia and Shu [121] proposed a CNN–Transformer hybrid architecture that enhances TransBTS with dual ViT layers and 3D CBAM attention in the encoder. While effective in capturing both local and global dependencies, the method required heavy computation and showed performance variability on unseen test data. Zeineldin et al. [337] proposed TransXAI, a hybrid CNN–Transformer model for glioma segmentation in multimodal MRI, emphasizing clinical interpretability through Grad-CAM visualizations. The model provides saliency maps that improved trust among neurosurgeons. The work highlights explainability as a critical step toward clinical adoption. Tariq et al. [290] proposed a hybrid framework integrating EfficientNetV2 and Vision Transformers for multiclass brain-tumor classification on MRI scans. By combining convolutional efficiency with transformer-based global context and further applying ensemble learning, the model achieved up to 96 % accuracy. Hossain et al. [102] proposed a hybrid ensemble transfer-learning model, combining VGG16, InceptionV3, and VGG19 for multiclass brain tumor classification, benchmarked against individual CNNs and Vision Transformers.

Taken together, the comparative analyses of CNN-, U-Net–, Transformer-, and hybrid-based transfer-learning approaches highlight the evolution of brain-tumor imaging research. Early CNN frameworks demonstrated the feasibility of transfer learning for classification tasks, while U-Net variants extended these ideas to volumetric segmentation with improved boundary delineation. The recent adoption of Transformer-based models has enabled more effective modeling of global dependencies, often surpassing CNN-only methods on benchmark datasets. Hybrid frameworks further demonstrate that combining multiple paradigms—such as CNN feature extractors with Transformer encoders, or GANs with transfer learning—can yield complementary benefits. This progression underscores a clear research trajectory, moving from handcrafted convolutional designs to more integrated and domain-adapted transfer-learning architectures for clinical decision support.

4.4 Comparative summary of reported studies

While Sections 4.3.1–4.3.4 presented detailed category-wise descriptions of transfer–learning approaches, this section consolidates the broader insights that emerge when comparing studies across CNN-, U-Net–, Transformer-, and hybrid-based methods.

A key observation is the evolution of transfer-learning applications in brain-tumor imaging. Early CNN-based models such as VGG, ResNet, and Inception primarily focused on feature extraction for classification tasks, achieving strong accuracy on Kaggle and institutional datasets. However, these studies often relied on relatively small datasets and lacked external clinical validation, which limited generalizability.

U-Net–based transfer learning extended these ideas to segmentation by incorporating pretrained encoders such as ResNet and EfficientNet. These approaches consistently demonstrated improved Dice similarity scores, particularly for whole-tumor delineation. Nevertheless, challenges persisted in segmenting enhancing regions, where class imbalance and limited training data remain critical bottlenecks.

Transformer-based transfer learning marked a significant methodological shift, enabling global context modeling through self-attention. Models such as Swin UNETR and ViT variants not only reported state-of-the-art Dice and Hausdorff metrics but also showed improved robustness to variations in MRI contrast. Their main limitation lies in high computational cost, with many studies requiring multi-GPU training and ensemble strategies to reach top leaderboard positions.

Hybrid strategies illustrate the latest stage in this progression, combining CNNs, Transformers, GANs, and optimization frameworks. These methods aim to balance local feature extraction with global modeling, while also addressing interpretability and dataset imbalance. GAN-based hybrids, for example, contribute sharper boundary delineation, while ensemble transfer learning frameworks leverage multiple pretrained backbones to achieve robust performance. Despite these innovations, hybrid approaches remain complex, with increased risk of overfitting and limited evaluation across diverse clinical cohorts.

Overall, the comparative evidence shows a steady improvement in reported performance metrics—from classification accuracies of around 90–95 % in early CNN transfer learning to Dice scores above 0.90 in Transformer-based methods and hybrid models. The most consistent benchmarking platform remains the BraTS challenge, which ensures comparability across studies, though many works continue to rely on smaller institutional datasets. The persistence of common gaps—such as lack of multi-institutional testing, limited explainability, and high computational requirements—indicates that future research should not only pursue higher accuracy but also focus on clinically viable deployment strategies.

4.5 Challenges and limitations

While transfer learning has significantly advanced brain-tumor detection and segmentation, several limitations remain.

1. The most transfer-learning pipelines rely on models pretrained on natural image datasets such as ImageNet, which are domain-mismatched compared to medical images. This can limit the transferability of higher-level features, especially for subtle tumor boundaries.
2. The dataset size and heterogeneity still pose challenges. MRI scans vary across scanners, institutions, and acquisition protocols, and pretrained models often fail to generalize without substantial fine-tuning. In particular, models optimized on public benchmarks like BraTS may not maintain accuracy when deployed on clinical datasets with different characteristics.
3. The class imbalance and the presence of small enhancing regions make segmentation tasks especially difficult. Although transfer learning improves initialization, it does not fully solve the imbalance problem, often leading to over-segmentation of larger tumor regions, while missing small lesions.
4. Computational and interpretability issues remain. State-of-the-art transfer-learning approaches often involve large architectures such as ResNet or Transformers, which demand high GPU resources and extensive fine-tuning. Moreover, the "black-box" nature of these models raises concerns in clinical use, where explainability and reliability are essential for decision support.

4.6 Conclusion and future directions

Transfer learning has become a cornerstone of modern approaches to brain-tumor detection and segmentation. By reusing knowledge from pretrained networks such as AlexNet, VGG, ResNet, DenseNet, EfficientNet, and Vision Transformers, researchers have achieved improved accuracy, faster convergence, and better generalization compared to training from scratch. Review of recent works demonstrates that CNN-based and U-Net–based architectures remain highly effective, while transformer-driven and hybrid frameworks are emerging as state-of-the-art strategies.

Looking ahead, several promising directions can address the following existing limitations:

- *Domain-specific pretraining:* Developing large-scale annotated medical image datasets for pretraining could reduce reliance on ImageNet and improve feature relevance.
- *Self-supervised and semi-supervised learning:* Leveraging unlabeled MRI data can help overcome annotation scarcity and enhance representation learning.
- *Model efficiency:* Lightweight architectures and knowledge distillation may enable clinically practical solutions with lower computational cost.

- *Explainability and trustworthiness:* Integrating attention maps, uncertainty estimation, and explainable AI tools will improve clinical acceptance of transfer-learning models.
- *Cross-domain generalization:* Future research should emphasize validation across multiple institutions, scanners, and patient populations to ensure robustness in real-world deployment.

In conclusion, transfer learning has already shown its potential to transform brain-tumor imaging, but further advances in domain-specific pretraining, efficiency, and interpretability will be critical to translating these methods from research prototypes to reliable clinical tools.

5 Diabetic retinopathy detection using deep learning

Abstract: Diabetic retinopathy (DR) remains one of the leading causes of preventable blindness worldwide, and early detection is essential for reducing vision loss. Traditional screening methods rely on manual examination of retinal fundus photographs, a process that is accurate but resource-intensive and difficult to scale. Recent advances in deep learning have enabled the development of automated systems capable of grading disease severity directly from retinal images. This chapter reviews the progress of deep learning approaches in DR detection, beginning with image-level classification and extending to lesion-level segmentation and explainability-based methods. Key public datasets are introduced, and representative studies are highlighted to show how convolutional networks, U-Net derivatives, transformers, and hybrid models have shaped the field. The discussion also addresses the challenges of generalization, clinical validation, and real-world deployment. While notable progress has been made, unresolved issues such as dataset bias, interpretability, and integration into healthcare systems remain critical. The chapter concludes with a forward-looking perspective on how more diverse datasets, explainable outputs, and equitable deployment strategies may shape the next generation of AI-based DR screening tools.

Keywords: Medical image segmentation, Retinal fundus photography, Lesion segmentation, Explainable AI, Automated screening, Biomedical image analysis

5.1 Introduction

Diabetic retinopathy (DR) is a common microvascular complication of diabetes mellitus and a leading cause of preventable vision loss among working-age adults. Global syntheses indicate that roughly one in five people with diabetes have some form of DR, with the absolute burden projected to rise substantially by 2045 as diabetes prevalence increases worldwide [292]. In the United States, recent multisource modeling estimated that 9.60 million people (26.43 % of those with diabetes) had DR in 2021, including 1.84 million with vision-threatening disease [165].

Traditional screening depends on manual grading of retinal fundus photographs by trained clinicians. While effective, this workflow is time-intensive and subject to interobserver variability, which limits scalability in population screening programs. Deep learning methods—particularly convolutional neural networks (CNNs)—have demonstrated high sensitivity and specificity for detecting referable DR directly from fundus images, establishing a foundation for automated triage and screening pathways [89]. Beyond accuracy, clinical validation has progressed to autonomous, point-of-care systems evaluated in prospective trials, illustrating the feasibility of deployment in primary care settings [5].

https://doi.org/10.1515/9783111389059-005

This chapter reviews deep learning approaches for DR detection and grading, spanning image-level classification and lesion-level segmentation to transformer-based and hybrid architectures. We also discuss dataset characteristics and biases, explainability and clinician trust, and practical considerations for integrating AI tools into real-world screening workflows.

5.2 Background on diabetic retinopathy and imaging

5.2.1 Pathophysiology and stages of DR

Diabetic retinopathy develops as a consequence of chronic hyperglycemia damaging retinal microvasculature, which leads to vision impairment and, in advanced cases, blindness. In the nonproliferative stage (NPDR), lesions such as microaneurysms, dot hemorrhages, and hard exudates are observed. As the disease progresses, venous beading and intraretinal microvascular abnormalities can occur, eventually leading to proliferative diabetic retinopathy (PDR). This advanced stage is characterized by pathological neovascularization, which increases the risk of vitreous hemorrhage, tractional retinal detachment, and irreversible vision loss [165, 292].

5.2.2 Imaging modalities for DR

The cornerstone of DR screening is retinal fundus photography, which captures 2D retinal views. Mydriatic devices require pupil dilation for high-quality images, while nonmydriatic cameras facilitate rapid imaging without dilation—ideal for tele-ophthalmology and mass screenings. Additionally, widefield fundus imaging (up to 200°) enables detection of peripheral lesions often missed in standard imaging [323].

Optical coherence tomography (OCT) is essential for diagnosing and monitoring diabetic macular edema (DME). OCT delivers high-resolution, cross-sectional images of the retina, allowing accurate retinal thickness measurements. Variants such as spectral-domain and swept-source OCT enhance image depth and acquisition speed [323]. Fluorescein angiography (FA) remains the clinical gold standard for visualizing vascular leakage, non-perfusion, and neovascularization in DR. However, its reliance on intravenous dye limits its practical use in large-scale screenings. A promising noninvasive technique is optical coherence tomography angiography (OCTA), which maps retinal microvasculature without contrast dye. OCTA shows strong potential for detecting ischemic changes and microvascular abnormalities in DR [268].

5.2.3 Publicly available datasets

Several public datasets have enabled benchmarking and reproducibility of deep learning methods for DR as shown in Table 5.1. These vary in size, annotation type, and clinical scope. These datasets have enabled comparative evaluation and development of robust DR models. EyePACS [312] offers extensive scale but suffers from imbalanced classes, while Messidor provides consistency and image quality. IDRiD [212] and DIARETDB1 [127] support lesion-focused tasks but are limited by size. The newer APTOS dataset [255] reflects real-world diversity, and MAPLES-DR [148] enriches Messidor with detailed anatomical labels, aiding interpretability-focused model development.

Table 5.1: Summary of widely used diabetic retinopathy datasets.

Dataset	Size (images)	Annotation Type	Notes / Challenges
EyePACS (Kaggle) [312]	~88,000	DR severity (0–4)	Largest public fundus dataset; substantial class imbalance; widely used in DL studies.
Messidor/ Messidor-2 [65]	1,200/1,748	DR grading	High-quality images, well-curated; a benchmark database frequently cited in clinical AI studies.
IDRiD [212]	516	Image-level grading + lesion segmentation	Annotated for lesions such as microaneurysms, hemorrhages, and exudates; limited sample size.
DIARETDB1 [127]	89	Lesion-level segmentation	Small but provides lesion-specific ground truth, useful for microaneurysm and exudate detection tasks.
APTOS 2019 (Kaggle) [255]	5,584	DR severity grading	Real-world variability in image quality; commonly used for robustness testing.
MAPLES-DR (Messidor extension) [148]	198	Pixel-level segmentation (10 biomarkers)	Provides detailed anatomical and pathological labels for explainable DR models.

5.3 Deep learning approaches for DR detection

Deep learning has become the cornerstone of modern automated diabetic retinopathy (DR) detection [89, 294]. Unlike traditional machine learning pipelines that rely on handcrafted features such as vessel morphology or lesion descriptors, deep neural networks are capable of learning hierarchical representations directly from raw retinal images. This paradigm shift has enabled significant improvements in both image-level classification of DR severity and lesion-level segmentation for clinical interpretability [5, 237]. Over the past decade, a wide variety of architectures—including convolutional neural networks (CNNs), encoder–decoder structures such as U-Net, transformer-based models, and hybrid ensembles—have been explored in the context of DR [63, 215, 303]. Each category brings unique advantages: CNNs excel in feature extraction, U-Nets in pixel-wise lesion localization, transformers in capturing global dependencies, and hybrid models in integrating complementary strengths. The following subsections review representative studies within each category, highlighting their methodological innovations, datasets, and reported performance.

5.3.1 Image-level DR grading

In some of the early work, CNNs were used just for classification.
- Pratt et al. [215]: a straightforward CNN for grading. Small datasets, results encouraging but nowhere near clinical use. More of a proof that "deep nets can work here. Big jump came with large training sets.
- Gulshan et al. [89]: trained on >100k fundus images. Validated on U.S. + Indian data. Model reached ophthalmologist-level sensitivity/specificity. This was the "wake-up call" paper. After this, people started believing DR grading with AI was realistic.
- Ting et al. [294]: tested on multiethnic Asian cohorts. Showed generalization across populations and devices. Important because it addressed a common criticism of Gulshan's system.

Regulatory milestone:
- Abramoff et al. [5]: IDx-DR. First FDA-cleared autonomous AI diagnostic tool. Prospective multicenter trial in U.S. primary care. Showed safe deployment outside tertiary eye hospitals. This was a big deal.

Then, the scope widened.
- Dai et al. [64]: DeepDR system with multiple subnetworks (quality check, lesion detection, grading). Multitask learning improved robustness.
- Dai et al. [63]: prediction of "progression time". Not just "what is the grade now?", but "when will it get worse?" Moves toward personalized screening intervals.

Population-scale deployment:
- Ruamviboonsuk et al. [236, 237]: Thailand national screening. Deep learning system compared with regional graders, then rolled out prospectively. Strong example of AI moving from lab to a real national program.

Notes: Image-level systems = scalable, efficient, validated at national scale. But = black box. They say "referable DR" but don't show lesions. If we look at Table 5.2, we can see the main studies that shaped image-level DR grading. It starts with small CNN trials and then moves toward very large datasets, FDA approval, and eventually nationwide screening work.

Table 5.2: Representative deep learning studies for image-level diabetic retinopathy grading.

Ref.	Method/System	Key Metric	Dataset	Strength	Limitations
[215]	Custom CNN with data augmentation	Acc = 75 %, Sens = 30 %, Spec = 95 %	EyePACS dataset (80,000 images)	Feasibility of CNNs for multiclass DR grading	Performance limited by class imbalance
[89]	Inception-v3 CNN	Sens = 0.97, Spec = 93.4 %	Messidor-2 (1,748 images)	Large-scale study	Trained/tested only on fundus images
[294]	Multitask deep learning	Sens = 90 %, Spec = 91 % (primary)	Singapore National DR Screening (112,648 images)	Robust validation	Complex system
[5]	Autonomous AI with image quality + diagnostic CNN modules	Sens = 87 %, Spec = 90 %	Prospective multicenter trial in 10 US primary care sites (n = 900 participants)	Rigorous prospective design; robust across sex, race	Lower sensitivity due to real-world imaging variability
[64]	Transfer learning–based multitask CNN	AUCs: 0.94 (mild), 0.95 (moderate), 0.96 (severe), 0.97 (PDR)	SIM cohort (Shanghai): 666,383 images	Lesion-aware segmentation, and DR grading	Single-ethnic development cohort (Chinese)
[63]	AI system predicting individualized time-to-progression of DR	AUCs (1–5 yrs) = 0.82–0.89	Pretraining: 717,308 images	System to predict patient-specific DR progression timelines	Developed mainly in Chinese cohort
[237]	Deep learning for real-time DR screening	Acc = 94 %, Sens = 91 %, Spec = 95 %, PPV = 79 %	15,270 fundus images	Real-time integration at point-of-care	Higher referral rates for ungradable images
[236]	Deep learning system retrospectively validated against human graders	AUC = 0.99; Sens = 97 %, Spec = 96 %	Thailand dataset: 25,326 gradable images	Demonstrated DL exceeded regional human graders' sensitivity	Reliance on single fundus camera type

5.3.2 Lesion-level detection and segmentation

We already saw that image-level models give a grade but no evidence. That annoyed clinicians. So, lesion-level work grew out of that gap.

The idea: "show me the microaneurysms, hemorrhages, exudates; don't just give me a label."

- Orlando et al. [199]: kind of a half step. Handcrafted features + shallow CNN. Random Forest to fuse. Reasonable, but tiny lesions often slipped through. Candidate extraction stage felt old fashioned

Later datasets (IDRiD, FGADR) gave pixel labels. That opened the door to proper segmentation nets.

- Ullah's SSMD-Net [303]: multiple decoders for each lesion type. They even used unlabeled EyePACS (semi-supervised). It generalized better, but training got messy. Too many tasks at once; small labeled pool still an issue.
- Li et al. [149]: attention blocks, lesion relations. Combined segmentation + grading. Useful because it outputs both lesion maps and severity. Strong results across DDR, IDRiD, APTOS. But imbalance—MAs and HEs weaker.
- Bian et al. [41]: multiscale attention + lesion perception. Good at subtle lesion details, worked across datasets. Downside: heavy training, hand-tuned losses.
- Xu et al. [326]: transformer with vessel priors. Better on small lesions. Outperformed CNNs. But relies on vessel annotations. That's a barrier in practice.

Notes: Lesion-level = more transparent, less scalable. Table 5.3 is basically a snapshot of how lesion-focused methods evolved. Early work mixed hand-designed features with CNNs, while later models tried semi-supervised training, and the most recent ones use attention and transformers for more transparent outputs.

Next step: severity grading with built-in explainability. Models not only "point to lesions" but also try to justify the overall stage in ways doctors accept.

5.3.3 Severity grading with explainability

Lesion-level approaches made AI outputs more transparent by pointing to specific abnormalities, but clinicians also wanted to know how those findings influenced the overall stage. This gave rise to severity grading with explainability, where models provide both the grade and a justification for that decision.

- Quellec et al. [220]: one of the early works. Used deep image mining to point out suspicious retinal areas. No pixel-level labels required—the model itself identified the patterns linked to DR. Heatmaps gave a first glimpse of how networks "see" lesions.

Table 5.3: Representative deep learning methods for lesion-level detection and segmentation in diabetic retinopathy.

Ref.	Method/System	Key Metric	Dataset	Strength	Limitations
[303]	Semi-supervised multi-decoder U-Net	Acc = 0.92, IoU = 0.847	IDRiD	Robust across datasets	Relatively small labeled sets for IDRiD
[149]	Deep attention network with lesion-aware relation block	DR grading accuracy = 87 %	APTOS	Provides interpretable outputs (heatmaps, lesion-level)	Modest Dice/IoU scores for small lesions (MAs, HEs)
[41]	Multiscale attention block + Lesion perception block	mAUPR = 67 %, mDice = 61;	DIARETDB1	Effectively captures multi-scale lesion features	Requires manual lesion-weight tuning for loss function
[326]	Prior-guided attention fusion Transformer	mAUPR = 0.67, mIoU = 0.33, mF1 = 0.66	IDRiD	Effectively integrates vessel priors into lesion segmentation	Dependent on pre-trained vessel masks
[199]	Hybrid ensemble for red-lesion detection	AUC = 0.89 (screening), 0.93 (referral)	MESSIDOR (image-level)	Strong results across lesion- and image-level tasks	Dependent on candidate-detection stage

- Wang and Yang [320]: proposed regression activation maps (RAM). Instead of just a grade, the system could show which parts of the fundus image supported the classification. A step closer to interpretable grading.
- Abràmoff et al. [5]: famous IDx-DR trial. Most often cited for being FDA-cleared, but also important here: The system provided lesion-level overlays. These visual cues helped physicians understand why the algorithm said "referable DR."
- Jiang et al. [122]: moved into Grad-CAM. Built a multi-label classifier for DR signs and attached saliency maps to each prediction. This gave fine-grained visual evidence for multiple disease indicators in one image.
- Sharma and Lalwani [255]: a recent study. Combined multiple deep nets and added an explainable AI layer. Not just better accuracy, but also region-level and decision-level explanations, aiming at real-world use.

Notes: Explainability work did not change the grading task itself, but it made the results more acceptable in clinics. From early CAMs to Grad-CAM and ensemble explainers, the trend is clear: prediction plus justification. Table 5.4 brings together key studies on explainable DR severity grading. It shows the shift from early CAM and RAM methods to Grad-CAM, FDA-approved systems, and recent ensemble approaches that focus on transparency alongside accuracy.

Image-level grading showed that CNNs could really work at scale, even reaching FDA approval. Lesion-level work then pushed for more transparency, with maps of microaneurysms, hemorrhages, and exudates that doctors could check. Explainability combined both ideas, keeping accuracy while adding reasons through heatmaps, Grad-CAM, or ensemble models. The overall direction is clear: not just grading DR, but grading it in a way that doctors can understand and trust.

5.3.4 Generalization and real-world deployment

One of the recurring hurdles for automated DR detection is that performance in the lab does not always carry over to the clinic. A model trained on tens of thousands of EyePACS images may achieve strong accuracy during internal testing, but the same system often underperforms when tested on images from a different hospital using another fundus camera or serving a population with different demographics and disease prevalence. This drop is expected—real-world data include variability in lighting, focus, and patient cooperation—and such factors are rarely captured in curated training sets.

Independent validation has been critical to addressing this gap. Ting et al. [294] demonstrated that models built in one population can lose accuracy when applied to multiethnic cohorts, underscoring the need for cross-population evaluation. Likewise, the Thailand national screening program showed that a deep learning system could outperform regional human graders, but it also uncovered practical barriers such as

Table 5.4: Representative studies on severity grading with explainability in diabetic retinopathy.

Ref.	Method/System	Key Metric	Dataset	Strength	Limitations
[220]	ConvNet with heatmap generation via backpropagation + hue constraint	AUC 0.95 (Kaggle), 0.94 (e-ophtha);	Kaggle DR, e-ophtha	No manual lesion labels needed; improved interpretability	Sensitivity to small lesions remained limited
[320]	CNN + RAM for both grading and lesion heatmaps	~86.2 % accuracy on Kaggle DR five-class grading	Kaggle EyePACS	Interpretable CAM-based method for DR; easy visualization	Heatmaps coarse; not fine-grained for tiny lesions
[5]	IDx-DR; lesion-specific detectors with interpretable overlays	Sens = 87 %, Spec = 90.7 %	Prospective study at 10 primary care sites ($n = 900$)	FDA-cleared autonomous AI; robust across populations	Strong dependence on image quality; no adaptive retraining
[122]	ResNet + multi-label classification + Grad-CAM	Sens = 93.9 %, Spec = 94.4 %	3,228 fundus images, five lesion labels	Interpretable Grad-CAM; reduced annotation effort	Grad-CAM maps coarse
[255]	Adaptive Gabor + modified U-Net + DenseNet	Acc = 99 % (DiaRetDB1), 98 % (APTOS 2019)	DiaRetDB1, APTOS 2019	Robust preprocessing; explainable outputs	Architectural complexity

ungradable images and the need to adjust referral thresholds. These experiences highlight that generalization is not only a technical challenge but also one tied to workflow design and health system policies.

Deployment further requires ongoing monitoring. Camera hardware changes, new image-acquisition protocols, and shifting disease distributions can gradually erode model performance. Ruamviboonsuk et al. [236, 237] emphasized the importance of retraining and real-time validation when deploying nationwide screening tools. Clinicians are also more likely to adopt systems that provide interpretable outputs, such as lesion overlays or heatmaps, a feature demonstrated in FDA-cleared tools like IDx-DR [5]. Finally, regulators and public health authorities remain central to large-scale adoption, ensuring that AI systems integrate with existing infrastructure rather than remaining isolated pilot projects.

5.4 Challenges and limitations

Despite notable progress, several obstacles remain before deep learning can be considered a mature solution for diabetic retinopathy (DR) screening.

- One major issue is dataset bias. Public datasets such as EyePACS or Messidor have been crucial for development, but they often reflect specific imaging devices or limited populations. When models trained on these data are tested on new cohorts, performance frequently drops, highlighting the limits of generalization [294].
- Another challenge is image quality. In real-world programs, a sizable fraction of retinal photographs are blurred, poorly illuminated, or obscured by artifacts. These "ungradable" images reduce model reliability and may increase unnecessary referrals, as observed in the Thailand national screening study [236, 237]. Handling such cases requires either prescreening modules or fallback strategies involving human graders.
- Interpretability also continues to limit adoption. While lesion maps and heatmaps have improved transparency [122, 220], many clinicians still question whether automated outputs can be fully trusted, especially in borderline cases. The regulatory process adds further complexity. Even FDA-cleared tools such as IDx-DR [5] had to undergo rigorous prospective validation before approval, and this level of evidence is not yet available for most systems.
- Cost and infrastructure form another bottleneck. Large-scale deployment depends on affordable imaging devices, integration with electronic health records, and sustainable funding models. In low-resource settings where the burden of DR is rising, these factors are particularly pressing [323]. Finally, ethical and legal concerns—such as patient data privacy, accountability for misdiagnosis, and lack of standardization—remain unresolved.

5.5 Conclusion and future directions

Deep learning has transformed the way diabetic retinopathy (DR) can be detected and graded. Research has progressed from early pilot studies with limited accuracy to large-scale validations that match and, in some cases, exceed human performance. National programs and approved autonomous systems demonstrate that AI-assisted screening is not only feasible but already beginning to shape clinical practice. At the same time, important gaps remain. Models still face difficulties with generalization across populations and imaging devices. Ungradable or poor-quality images continue to reduce efficiency in large screening programs. Questions of interpretability persist, and regulatory, economic, and ethical barriers remain significant.

Looking ahead, several directions are likely to define the next phase of this field. Building more diverse and representative datasets will help ensure robustness. Approaches such as domain adaptation, federated learning, and continuous retraining can improve reliability when models encounter new environments. Greater emphasis on explainability will make these systems more acceptable to clinicians, while tighter integration with healthcare infrastructure will enable real-world scalability. Finally, lightweight and accessible solutions should be prioritized to extend screening to underserved regions where the burden of diabetes is rapidly increasing.

In conclusion, AI-based DR detection is moving from laboratory research to practical deployment. The focus now must shift from simply achieving high accuracy to ensuring that these systems are generalizable, transparent, and sustainable in real-world healthcare. Addressing these priorities will allow automated screening to play a meaningful role in reducing preventable vision loss worldwide.

6 A deep learning-based solution for automated melanoma detection

Abstract: Melanoma remains one of the most dangerous skin cancers, and timely identification is essential to improve patient outcomes. This chapter presents a structured view of recent artificial-intelligence approaches for analyzing dermoscopic images of pigmented lesions. After introducing the clinical context, public datasets, and widely used evaluation metrics, the discussion centers on three representative methods: MelaNet, which couples CycleGAN-based data augmentation with a VGG classifier; Seq-Diff-Net, a two-stream network that captures spatial and temporal lesion changes; and CKDNet, which integrates contextual knowledge diffusion for simultaneous segmentation and classification. A comparative study highlights their relative strengths and remaining limitations. The chapter concludes with open research issues and practical directions aimed at building accurate, interpretable, and lightweight systems for early melanoma screening.

Keywords: Melanoma detection, Dermoscopic imaging, Deep learning architectures, Skin lesion segmentation, Classification

6.1 Introduction

Skin cancer is among the most common malignancies worldwide, and melanoma is its most aggressive form. Although melanoma represents only about 1 % of skin cancers, it accounts for the majority of skin cancer-related deaths [324]. According to the World Health Organization, approximately 325,000 new melanoma cases and 57,000 deaths were reported in 2020, and the incidence continues to rise globally. Early diagnosis is therefore critical: Studies indicate that detecting melanoma at an initial stage can achieve a five-year survival rate above 99 % [178].

Traditional diagnostic techniques include visual inspection with the ABCDE rule (Asymmetry, Border irregularity, Colour variation, Diameter, Evolving features), dermoscopy, and histopathological examination. While dermoscopy improves detection accuracy by 10–20 % over naked-eye inspection, it requires expert training and is prone to interobserver variability. Advanced optical imaging methods such as reflectance confocal microscopy (RCM) and multiphoton tomography (MPT) offer high accuracy but are expensive and not widely available, especially in remote or low-resource settings. As a result, patients in rural areas often lack access to timely screening, and false positives can lead to unnecessary biopsies.

Recent advances in *artificial intelligence* (AI) and computer-aided diagnosis (CAD) provide new opportunities to address these challenges. Convolutional neural networks (CNNs) and related deep-learning architectures have shown superior performance to

https://doi.org/10.1515/9783111389059-006

expert dermatologists in several studies, particularly when trained on large dermoscopic datasets [75]. These systems can assist clinicians in lesion segmentation and classification, improving accuracy and enabling wider deployment via teledermatology platforms. However, barriers remain, including limited availability of diverse, well-annotated datasets, the need for lightweight models for mobile devices, and concerns regarding interpretability and clinical integration.

Motivation and Scope

This chapter presents a comprehensive overview of AI-based approaches for melanoma detection using dermoscopic images. We first review classical computer-vision and deep-learning techniques, followed by detailed case studies of advanced architectures such as CycleGAN–VGG (MelaNet), Seq-Diff-Net, and CKDNet. Datasets, performance metrics, and comparative analyses are discussed, along with practical challenges such as class imbalance, boundary uncertainty, and resource constraints. Finally, we outline future research directions, including explainable and lightweight AI models, multimodal fusion, and clinical deployment pathways.

6.2 Background and related work

Automated melanoma detection has evolved from simple image-processing techniques to advanced deep-learning frameworks. Early approaches relied on handcrafted features and rule-based decision systems, whereas recent research leverages convolutional and generative models to improve accuracy and generalizability. This section reviews major families of methods, highlighting their strengths, limitations, and open challenges.

6.2.1 Classical computer-vision approaches

Traditional computer-vision methods analyze lesion images through handcrafted feature extraction followed by classification. Techniques such as adaptive thresholding, region growing, active contours, and support vector machines (SVM) have been used for segmentation and lesion classification [46, 60]. Although these approaches improved over basic visual inspection, they often fail in the presence of artefacts such as hairs, shadows, or low contrast. They also depend on careful parameter tuning and offer limited scalability to large datasets.

6.2.2 Deep learning architectures

The introduction of deep convolutional neural networks (CNNs) revolutionized medical image analysis. Architectures such as Fully Convolutional Networks (FCN) and U-Net enabled end-to-end semantic segmentation, while residual and densely connected networks (ResNet, DenseNet) enhanced feature reuse and mitigated vanishing gradients [98, 232]. Lightweight models (e. g., MobileNet, LCASA-Net) have been explored for deployment on mobile or edge devices, though compression techniques may degrade accuracy if not tuned carefully. Advanced designs including pyramid pooling and residual attention modules capture multiscale contextual information, further boosting segmentation and classification performance.

6.2.3 GAN-based methods

Generative Adversarial Networks (GANs) have been introduced to tackle data scarcity and class imbalance. Variants such as CycleGAN, GAN-based U-Net, and dual-discriminator frameworks generate synthetic lesions or refine segmentation boundaries [117, 147]. By enriching underrepresented classes, these models reduce overfitting and help algorithms generalise to unseen cases. However, adversarial training can be computationally expensive, and careful validation is needed to ensure synthetic images do not introduce artefacts or bias.

6.2.4 Attention mechanisms

Attention modules improve the discriminative power of CNNs by focusing on salient image regions. Approaches include soft self-gated attention, attention gates, and channel–spatial attention blocks [197, 244]. These mechanisms enhance lesion localization without substantially increasing model complexity, making them attractive for real-time and resource-constrained applications.

6.2.5 Research gaps and challenges

Despite rapid progress, several gaps remain in automated melanoma detection as follows:

- *Lesion boundary complexity:* Irregular, blurred borders and artefacts such as hair or illumination still degrade segmentation accuracy.
- *Dataset imbalance and bias:* Malignant samples are underrepresented in most public datasets, and images from darker skin tones remain limited.

- *Generalisation:* Models trained on specific datasets may lose accuracy when deployed across new devices, lighting conditions, or populations.
- *Resource constraints:* Many high-performing models require significant computational resources, limiting their use in teledermatology or mobile platforms.
- *Interpretability:* Deep networks often behave as “black boxes”, hindering clinician trust and regulatory adoption.

6.3 Datasets and evaluation protocols

Robust datasets and clear evaluation protocols are essential for developing reliable AI models for melanoma detection. They provide the foundation for algorithm training, benchmarking, and fair comparison between competing methods.

6.3.1 Overview of benchmark datasets

ISIC archive and challenges

The International Skin Imaging Collaboration (ISIC) has released several benchmark datasets for skin lesion analysis:
- *ISIC 2016*: Early dataset of ~900 dermoscopic images, primarily used for lesion segmentation [58].
- *ISIC 2018*: Contains 2,594 training and 1,000 test images with high-quality segmentation masks, partly derived from the HAM10000 dataset [58].
- *ISIC 2019*: A large-scale collection of 25,331 training and 8,238 test images. Metadata such as patient age, sex, and anatomical site enable more detailed analyses [84].

HAM10000

The *Human Against Machine with 10,000 training images* (HAM10000) dataset includes 10,015 dermoscopic images gathered from multiple sources and acquisition settings [301]. It is widely used for classification tasks and serves as the basis for several ISIC challenges.

PH2

PH2 comprises 200 dermoscopic images (40 melanomas, 80 atypical nevi, and 80 common nevi) at a resolution of 768 × 560 pixels. Each image is accompanied by expert annotations, making PH2 a valuable benchmark for segmentation studies [176].

DermoFit

The DermoFit Image Library from the University of Edinburgh offers approximately 1,300 high-quality RGB images across ten lesion categories, including 76 melanomas and several benign classes [28]. Its controlled acquisition conditions support algorithmic evaluation, though the dataset remains relatively small.

Other resources, such as BCN20000 and private institutional archives, are also used in research, although public access is often restricted.

6.3.2 Dataset bias and representation

While these datasets underpin most research on melanoma detection, they present several challenges:

- *Class imbalance:* Malignant lesions are consistently underrepresented compared to benign ones, which can bias model training.
- *Skin-tone diversity:* Images from darker skin tones and rarer lesion subtypes remain limited.
- *Acquisition gap:* Many images are captured under controlled laboratory conditions, which may not reflect real-world clinical practice.

Addressing these limitations requires curated datasets that include diverse demographics, imaging devices, and lesion types.

6.3.3 Evaluation metrics

Accurate assessment of segmentation and classification models relies on well-defined metrics.

6.3.3.1 Segmentation metrics

The most widely used measures are the Dice Similarity Coefficient (DSC) and Intersection over Union (IoU):

$$\mathrm{DSC} = \frac{2|P \cap G|}{|P| + |G|}, \tag{6.1}$$

$$\mathrm{IoU} = \frac{|P \cap G|}{|P \cup G|}, \tag{6.2}$$

where P and G denote the predicted and ground-truth lesion regions, respectively [69].

6.3.3.2 Classification metrics

For lesion classification, accuracy, sensitivity (recall), specificity, and precision are typically reported:

$$\text{Accuracy} = \frac{\text{TP} + \text{TN}}{\text{TP} + \text{TN} + \text{FP} + \text{FN}}, \tag{6.3}$$

$$\text{Sensitivity} = \frac{\text{TP}}{\text{TP} + \text{FN}}, \tag{6.4}$$

$$\text{Specificity} = \frac{\text{TN}}{\text{TN} + \text{FP}}, \tag{6.5}$$

$$\text{Precision} = \frac{\text{TP}}{\text{TP} + \text{FP}}, \tag{6.6}$$

where TP, TN, FP, and FN denote true positives, true negatives, false positives, and false negatives.

The Area Under the ROC Curve (AUC) is widely used to summarize classifier performance across thresholds:

$$\text{AUC} = \int_0^1 \text{TPR}(\text{FPR})\, d(\text{FPR}), \tag{6.7}$$

with TPR and FPR representing the true-positive and false-positive rates, respectively.

6.3.4 Summary

Public datasets such as ISIC, HAM10000, PH2, and DermoFit have enabled major advances in AI-based melanoma detection. Nevertheless, gaps in malignant case representation, limited inclusion of darker skin tones, and discrepancies between curated datasets and real-world images remain obstacles. Future work should emphasize the creation of large, demographically balanced datasets and encourage rigorous cross-dataset validation.

6.4 Case studies: advanced architectures

6.4.1 CycleGAN-VGG (MelaNet): addressing class imbalance

Class imbalance is a persistent issue in melanoma datasets, where malignant lesions are typically underrepresented. To mitigate this, *MelaNet* integrates CycleGAN-based image synthesis with a VGG-style classifier. The approach enriches the minority class through realistic augmentation, enabling more balanced training and improved sensitivity to malignant cases [264, 356].

Method

MelaNet employs a Cycle-Consistent Generative Adversarial Network (CycleGAN) to generate synthetic melanoma images. Two generators $G : X \to Y$ and $F : Y \to X$ map between benign (X) and malignant (Y) lesion domains. The adversarial objective for G and its discriminator D_Y is

$$\mathcal{L}_{\mathrm{GAN}}(G, D_Y, X, Y) = \mathbb{E}_{y \sim p_{\mathrm{data}}(y)}[\log D_Y(y)] + \mathbb{E}_{x \sim p_{\mathrm{data}}(x)}[\log(1 - D_Y(G(x)))], \tag{6.8}$$

with a similar term for F and D_X. A cycle-consistency loss preserves lesion structure:

$$\mathcal{L}_{\mathrm{cyc}}(G, F) = \mathbb{E}_{x \sim p(x)}[\|F(G(x)) - x\|_1] + \mathbb{E}_{y \sim p(y)}[\|G(F(y)) - y\|_1]. \tag{6.9}$$

The enriched dataset (real and synthetic images) is then used to train a VGG-based convolutional network for melanoma classification. Feature extraction relies on stacked convolutional layers with ReLU activations and max-pooling, followed by fully connected layers and a softmax output. As shown in Fig. 6.1, the proposed MelaNet model combines CycleGAN-based augmentation with a VGG backbone for melanoma classification.

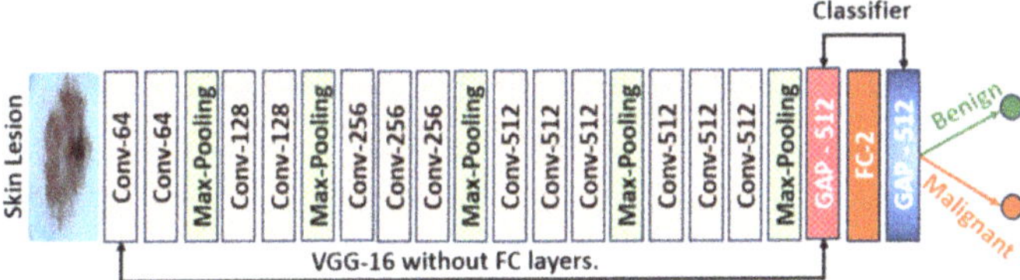

Figure 6.1: Architecture of MelaNet: a CycleGAN-based augmentation pipeline combined with a VGG classifier for melanoma detection (adapted from [344]).

Dataset and experimental setup

Experiments used the ISIC 2018 dataset, following its official training–validation split. Images were resized to 256 × 256 pixels and normalized before training. CycleGAN was trained with a learning rate of 2×10^{-4}, batch size of 4, and 200 epochs, producing approximately 3,000 synthetic melanoma images. The classifier used stochastic gradient descent with momentum, an initial learning rate of 10^{-3}, and early stopping based on validation accuracy.

Results and discussion

Table 6.1 compares MelaNet with a baseline VGG network trained without augmentation. The proposed model achieved an AUC of 0.942 and sensitivity of 0.903, exceeding the baseline by over 5 % in sensitivity, while retaining high specificity.

Advantages and limitations

Integrating GAN-based augmentation with a VGG classifier improves sensitivity to malignant cases and preserves lesion structure via cycle-consistency constraints. The approach is compatible with standard CNN pipelines; however, adversarial training can be computationally demanding, and poor convergence may produce unrealistic samples. Future work should explore lightweight GAN variants and assess the clinical acceptance of synthetic images.

6.4.2 Seq-Diff-Net: spatio-tTemporal modelling of lesion evolution

Early detection of changes in a skin lesion over time can be critical for recognizing melanoma at an incipient stage. *Seq-Diff-Net* was proposed to explicitly capture these temporal dynamics by modelling the evolution of dermoscopic images across follow-up visits. The network integrates a spatial feature extractor with a temporal differencing module, enabling it to focus on progressive alterations in colour, shape, and boundary [1].

Method

The framework consists of a convolutional backbone for spatial encoding and a sequential differencing layer that aggregates features from consecutive images. Given a series of dermoscopic images $\{I_t\}$, spatial features F_t are obtained through a shared encoder. A difference operator emphasizes local changes:

$$D_t = F_t - F_{t-1}, \tag{6.10}$$

and these difference maps are fed to a recurrent aggregation module (e. g., GRU) to model temporal patterns. The final representation is passed to a classification head for melanoma prediction.

Dataset and experimental setup

Experiments were carried out using a subset of the ISIC archive containing longitudinal image series of pigmented lesions. Images were resized to 224×224 pixels, normalized, and arranged chronologically for each patient. Training employed a learning rate of 5×10^{-4} with Adam optimizer and early stopping based on validation AUC. Data augmentation included random flips, rotations, and slight brightness shifts to mimic variations in acquisition.

Results and discussion

Seq-Diff-Net achieved an AUC of 0.935 and a sensitivity of 0.888, outperforming baseline CNNs that analyzed each image independently. Temporal differencing allowed the model to detect subtle pigment changes and edge irregularities that might be overlooked in static analyses.

Advantages and limitations

By explicitly modelling lesion evolution, Seq-Diff-Net enhances early melanoma recognition and reduces missed detections for slowly developing cases. However, the method requires sequential image data, which may not always be available in practice. Further work could explore self-supervised pretraining or synthetic progression modelling to mitigate data scarcity. The temporal processing pipeline of Seq-Diff-Net is illustrated in Fig. 6.2, highlighting its dual-stream design for capturing lesion evolution.

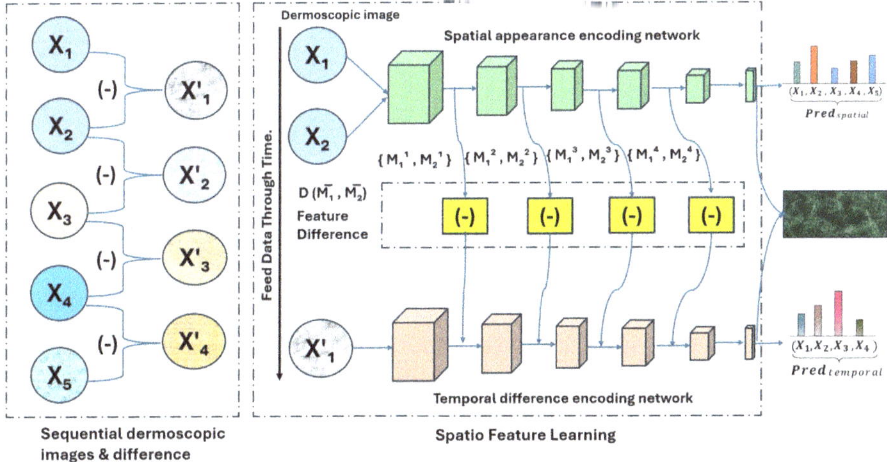

Figure 6.2: Seq-Diff-Net architecture for modelling spatial and temporal lesion evolution in dermoscopic images (adapted from [1]).

6.4.3 CKDNet: joint segmentation and classification via knowledge diffusion

Accurate melanoma assessment often requires both precise lesion delineation and reliable classification. *CKDNet* (*Contextual Knowledge Diffusion Network*) was proposed to unify these tasks, encouraging information sharing between segmentation and diagnosis modules. The goal is to leverage contextual cues from the predicted mask to guide classification, while classification feedback refines the segmentation boundary [2].

Method

CKDNet adopts a dual-branch design comprising a segmentation backbone and a classification head, connected through a knowledge-diffusion module. The segmentation path follows a U-Net-like encoder–decoder that outputs a lesion probability map. Features from intermediate decoder layers are propagated to the classifier, enabling it to focus on relevant regions. Conversely, gradients from the classification branch inform seg-

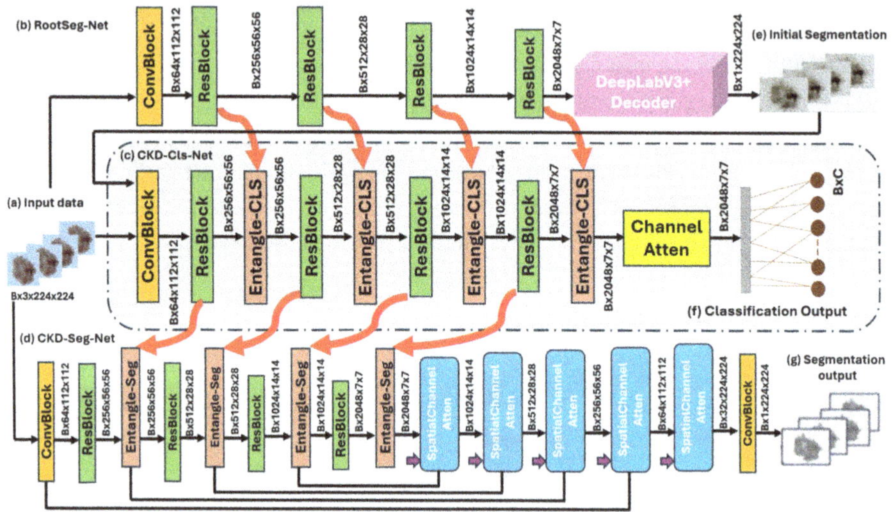

Figure 6.3: Overview of CKDNet, a contextual knowledge diffusion network for joint segmentation and classification of melanoma lesions (adapted from [2]).

mentation refinement. Figure 6.3 presents the CKDNet architecture, which integrates contextual knowledge diffusion to jointly handle segmentation and classification.

The overall objective combines a weighted cross-entropy loss $\mathcal{L}_{\text{ce}}$, Dice loss $\mathcal{L}_{\text{dice}}$, and a focal term to handle class imbalance:

$$\mathcal{L}_{\text{seg}} = \alpha\,\mathcal{L}_{\text{ce}} + \beta\,\mathcal{L}_{\text{dice}} + \gamma\,\mathcal{L}_{\text{focal}}, \tag{6.11}$$

while the classification branch minimizes:

$$\mathcal{L}_{\text{cls}} = -\sum_i y_i \log(\hat{y}_i), \tag{6.12}$$

and the final loss is a weighted sum:

$$\mathcal{L}_{\text{total}} = \mathcal{L}_{\text{seg}} + \lambda\,\mathcal{L}_{\text{cls}}. \tag{6.13}$$

Dataset and experimental setup

CKDNet was evaluated on the ISIC 2019 dataset, which offers over 25,000 dermoscopic images spanning eight diagnostic categories. All images were resized to 256 × 256 pixels, and lesion masks were used where available. Training used Adam optimizer with a learning rate of 10^{-4} and early stopping based on validation Dice. Data augmentation (random scaling, elastic deformations, and color jitter) helped improve robustness to acquisition variability.

Results and discussion

As summarized in Table 6.1, CKDNet achieved a Dice coefficient of 0.917 for segmentation and an AUC of 0.948 for classification, exceeding single-task baselines by a notable margin. Knowledge diffusion improved boundary localisation and reduced false positives near lesion edges.

Advantages and limitations

CKDNet demonstrates the benefit of joint optimization for segmentation and classification in melanoma analysis. By sharing contextual knowledge, the network improves both boundary accuracy and diagnostic sensitivity. However, its coupled training increases memory requirements, and performance may decline if lesion masks are unavailable during training. Future work should investigate semi-supervised learning or lightweight backbones to extend CKDNet to resource-constrained environments.

6.5 Comparative analysis of advanced architectures

Evaluating the performance of multiple architectures on common benchmarks provides insight into their relative strengths and limitations. Table 6.1 summarizes the results obtained by MelaNet, Seq-Diff-Net, and CKDNet on the ISIC 2018 and ISIC 2019 datasets, alongside baseline models such as standard CNNs and U-Net variants.

Table 6.1: Performance comparison of advanced architectures for melanoma detection.

Model	Dataset	AUC	Sensitivity	Dice / IoU
VGG (baseline)	ISIC 2018	0.892	0.848	–
MelaNet (CycleGAN–VGG)	ISIC 2018	0.942	0.903	–
Seq-Diff-Net	ISIC 2018 (longitudinal)	0.935	0.888	–
U-Net (baseline)	ISIC 2019	–	–	0.881 / 0.804
CKDNet	ISIC 2019	0.948	0.915	0.917 / 0.835

The results show that *MelaNet* achieves the largest gain in sensitivity compared to a plain VGG classifier, thanks to CycleGAN-based augmentation that addresses class imbalance. *Seq-Diff-Net* performs particularly well in scenarios where lesion evolution over time is available, demonstrating the value of temporal modelling for early detection. *CKDNet* provides the most balanced performance overall, delivering strong segmentation (Dice 0.917) and classification (AUC 0.948) within a single framework.

Overall, the analysis indicates that data augmentation (MelaNet), temporal reasoning (Seq-Diff-Net), and joint optimization (CKDNet) each contribute to improved melanoma detection. The choice of method should depend on the availability of longitudinal data, computational resources, and whether accurate lesion segmentation is required alongside classification.

The comparative results above emphasize how different design choices affect performance. Nevertheless, deploying these systems in practice introduces further obstacles, discussed next.

6.6 Challenges and research gaps

Despite rapid progress in AI-driven melanoma detection, several technical and practical barriers limit deployment in routine clinical care. Understanding these challenges is essential for guiding future research.

Data scarcity and class imbalance. Most publicly available datasets contain far fewer malignant than benign samples, and many lesion categories are underrepresented. This imbalance biases training and reduces sensitivity to rare melanomas. Synthetic augmentation (e. g., GAN-based) helps but requires careful validation to avoid artefacts or bias [154].

Dataset bias and skin-tone diversity. Images of darker skin tones, paediatric cases, or unusual subtypes are sparse in standard archives. Models trained primarily on fair-skinned adults may not generalize to other populations, raising fairness and safety concerns.

Segmentation difficulties and image artefacts. Hairs, shadows, reflections, and blurred lesion boundaries continue to challenge automated segmentation. Robust pre-processing and boundary-aware loss functions can mitigate some errors, but small or low-contrast lesions remain difficult.

Generalization across devices and environments. Differences in camera type, resolution, illumination, and acquisition protocol often cause performance drops when models are deployed outside the dataset they were trained on. Cross-dataset evaluation and domain adaptation techniques remain underused.

Resource and latency constraints. High-capacity architectures such as GANs or knowledge-diffusion networks may exceed the computational budget of mobile devices or teledermatology platforms. Lightweight designs, pruning, and quantization strategies require further research to maintain accuracy while lowering memory and inference cost.

Interpretability and trust. Many deep networks still behave as "black boxes", providing little insight into their decisions. Saliency maps and attention mechanisms improve transparency, but standardized interpretability tools and user studies with clinicians are needed for adoption.

Ethical, privacy, and regulatory considerations. Use of patient data raises privacy concerns, especially when models are trained on multi-institutional datasets. Federated learning and privacy-preserving techniques can address this but are not yet routine. Clear regulatory pathways for AI-based diagnostic tools remain to be defined.

Overall, these challenges suggest that future systems should emphasize data diversity, strong validation across settings, explainability, and efficient architectures suited for real-world clinical environments.

6.7 Future directions

Ongoing advances in computer vision, generative modelling and mobile computing offer opportunities to overcome current barriers in AI-based melanoma detection. Several promising avenues are outlined below.

Multimodal data fusion. Combining dermoscopic images with clinical data such as patient demographics, lesion history, or genetic risk factors could improve predictive accuracy and personalize screening [29]. Fusion architectures may integrate tabular and image features through joint attention or graph-based reasoning.

Explainable and trustworthy AI. Interpretability tools need to evolve from post-hoc visualization to intrinsic explainability. Embedding attention modules, prototype learning, or concept attribution directly into models can increase transparency and clinician confidence. User-centered studies should evaluate how explanations affect diagnostic decisions.

Lightweight and edge-ready models. Deploying melanoma detection to mobile or point-of-care devices requires compact yet accurate architectures. Model pruning, quantization, and knowledge distillation can reduce complexity, while hybrid cloud–edge pipelines may balance latency and privacy.

Federated and privacy-preserving learning. Collaborative training across institutions without sharing raw images can enlarge datasets while protecting patient confidentiality. Federated learning, combined with differential privacy or secure aggregation, could help build globally robust models [172].

Synthetic and self-supervised data strategies. Generative models and self-supervised pretraining can alleviate the dependence on large, fully labelled datasets. Future work should evaluate synthetic lesions for diversity and realism and explore contrastive learning to leverage unlabeled archives.

Clinical integration and longitudinal validation. Real-world adoption depends on embedding algorithms into clinical workflows and assessing their impact through prospective studies. Pilot programmes with teledermatology services and community screening initiatives can provide feedback on usability, bias, and patient outcomes.

Taken together, these directions point toward systems that are accurate, interpretable, and accessible, capable of supporting early melanoma detection on a global scale.

6.8 Conclusion

Artificial intelligence has shown substantial promise for supporting the early detection of melanoma, offering performance that rivals or exceeds expert assessment on benchmark datasets. This chapter reviewed the evolution of automated methods, from classical image-processing techniques to contemporary deep learning solutions. We discussed three advanced architectures—MelaNet, Seq-Diff-Net, and CKDNet—highlighting how data augmentation, temporal reasoning, and joint optimization can address key challenges in melanoma analysis. A comparative study across public datasets demonstrated that each approach contributes unique strengths: MelaNet improves sensitivity by enriching minority classes; Seq-Diff-Net leverages sequential images to track lesion evolution; and CKDNet enhances both segmentation and classification through knowledge diffusion. At the same time, limitations related to data scarcity, bias, and computational demands underscore the need for carefully curated datasets, rigorous validation, and efficient model design. Looking forward, progress will depend on combining rich and diverse data resources with explainable, privacy-aware, and lightweight algorithms that can be integrated into clinical workflows. Such systems hold the potential to expand access to accurate melanoma screening and reduce the burden of skin cancer worldwide.

7 Automated anterior cruciate ligament (ACL) injury detection by magnetic resonance imaging

Abstract: Anterior cruciate ligament (ACL) injuries are frequently observed in athletes engaged in high-intensity or pivoting sports and often lead to reduced knee stability, restricted mobility, and impaired functional performance. Accurate and timely diagnosis—whether for partial damage or complete rupture—is critical for guiding appropriate treatment strategies and ensuring effective rehabilitation. Magnetic Resonance Imaging (MRI) continues to serve as the preferred method for assessing ACL integrity, although manual image interpretation can be both time-consuming and prone to variability among radiologists. In recent years, artificial intelligence (AI), particularly deep learning techniques such as convolutional neural networks (CNNs), has shown considerable potential in automating the detection of ACL injuries and enhancing diagnostic precision. This chapter highlights the anatomical structure of the ACL, outlines typical mechanisms of injury, and reviews the application of MRI-based diagnostic approaches, discussing their advantages as well as the limitations that persist in clinical practice. This chapter examines both semi-automated and fully automated approaches for detecting anterior cruciate ligament (ACL) injuries. Semi-automated methods support radiologists by segmenting MRI scans and extracting relevant features, with the final diagnosis relying on clinician interpretation. Advances in artificial intelligence have enabled the development of fully automated diagnostic systems, most of which are based on convolutional neural networks (CNNs), designed to improve the efficiency and accuracy of ACL injury detection without direct clinician input. While these systems show considerable potential for streamlining workflows and aiding clinical decision-making, challenges persist, particularly regarding the interpretability of model outputs and the limited diversity of available training datasets. Performance is commonly reported using sensitivity, specificity, and the area under the receiver operating characteristic curve (AUC). Together, these metrics speak to both diagnostic accuracy and a model's ability to generalize across settings. The chapter also foregrounds the growing role of explainable AI (XAI) and hybrid approaches that pair data-driven algorithms with clinical expertise, aiming to deliver high diagnostic performance without sacrificing transparency or trust in practice.

Keywords: Anterior cruciate ligament (ACL) injury detection, Magnetic resonance imaging (MRI), Biomedical image analysis

7.1 Introduction

The anterior cruciate ligament (ACL) is a key stabilizing structure of the knee, preventing excessive forward movement and rotation of the tibia in relation to the femur. Injuries

https://doi.org/10.1515/9783111389059-007

to the ACL are common in sports that require sudden deceleration, rapid changes in direction, or unbalanced landings, such as soccer, basketball, football, and skiing. These injuries span a spectrum from mild sprains or partial tears to complete ruptures, with complete tears often necessitating surgical reconstruction to restore joint stability. Even partial tears may predispose individuals to chronic instability or early-onset osteoarthritis if left untreated [15]. Bedside maneuvers such as the Lachman and pivot-shift tests can be informative, but in the immediate post-injury setting their reliability often falls off because pain, guarding, and joint effusion limit examination.

MRI remains the modality of choice for evaluating ACL injury, offering high-resolution views of the ligament and adjacent anatomy, including cartilage and menisci [257]. Yet manual reading is time-consuming and subject to interobserver variation—particularly for partial tears and subtle findings—underscoring the need for consistent, objective tools that can support decision-making and improve patient outcomes.

7.1.1 Importance of ACL injury detection

Accurate, timely identification of anterior cruciate ligament (ACL) injury is central to selecting appropriate treatment and rehabilitation. Early diagnosis helps determine whether surgical reconstruction or structured conservative care is most suitable. By contrast, delays or errors can expose the knee to secondary harm—most notably meniscal tears and chondral damage—and increase the long-term risk of post-traumatic osteoarthritis (PTOA) [164]. For elite athletes, time to diagnosis often governs the safety and timing of return to play, which raises the premium on precision. In this context, recent advances in artificial intelligence (AI) have enabled automated MRI-based approaches for detecting ACL injury. Deep learning models, particularly convolutional neural networks (CNNs), have shown strong capabilities in interpreting MRI scans to identify ACL tears. When trained on large, well-annotated datasets, these models can capture complex injury patterns with high accuracy, which can speed up diagnosis and reduce the likelihood of human error [208].

7.1.2 Challenges involved in automated ACL injury detection

Although AI has enabled substantial progress in medical imaging, several challenges hinder the widespread clinical adoption of automated ACL injury detection. A primary limitation is the significant variability in MRI acquisition protocols across institutions; in particular, differences in scanner manufacturer, magnetic field strength, sequence parameters, image resolution, and contrast settings can all lead to inconsistent image characteristics. As a result, deep learning models developed in a single-center often fail to generalize to external data unless robust harmonization, data augmentation, or domain adaptation strategies are applied [189]. Detecting partial anterior cruciate ligament (ACL) tears remains a complex and clinically significant challenge. These injuries often manifest as subtle disruptions in ligament structure, which can be difficult to distinguish from intact tissue on MRI. A recent study by Tokgoz et al. (2024) reported that measuring the posterior cruciate ligament buckling angle provided detection sensitivities of approximately 87 % and specificities of around 90 % for partial ACL tears—highlighting the need for precise imaging techniques and tailored AI models trained on rich datasets that include such subtler injury patterns [296].

To enhance the accuracy of detecting partial tears, researchers have explored multi-view MRI data integration, combining information from sagittal, coronal, and axial views to improve diagnostic precision.

Interpretability is a prerequisite for clinical use of artificial intelligence (AI): Clinicians need to understand and trust model outputs before they can inform care. Although deep models such as convolutional neural networks (CNNs) often achieve strong accuracy, their internal logic is opaque, which makes the basis of a prediction hard to interrogate. This opacity remains a major barrier to adoption.

Explainable AI (XAI) methods, e. g., Grad-CAM, saliency maps, and concept activation mapping, aim to surface the image evidence that most influenced a decision, improving transparency and, in turn, clinician confidence. However, visual explanations must be clinically meaningful. Rigorous validation and close involvement of healthcare professionals are essential to ensure that highlighted regions and concepts align with medical reasoning and can be integrated smoothly into day-to-day workflows [209].

7.2 Anatomical structure of the ACL

7.2.1 Anatomy of the anterior cruciate ligament

Acting as the main restraint to anterior tibial displacement under the femur, the anterior cruciate ligament (ACL) contributes substantially to knee stability. It originates from the lateral femoral condyle and inserts on the anterior intercondylar area of the tibia; anatomically, it lies intra-articular yet extra-synovial [77]. Functionally, the ligament is most taxed during cutting, pivoting, and rapid changes of direction.

Classically, the ACL comprises two bundles with complementary roles: the anteromedial (AM) and the posterolateral (PL). Across the arc of motion they share load but contribute differently—AM fibers are more engaged in controlling rotation when the knee is flexed, whereas PL fibers provide greater restraint near full extension [14]. The tissue itself is dense and fibrous and tolerates substantial tensile forces, yet it is vulnerable during abrupt deceleration or pivoting maneuvers, circumstances that raise injury risk [171].

The ligament is not purely structural; embedded mechanoreceptors provide proprioceptive feedback to the central nervous system. This sensory feedback supports coordinated muscular activation around the joint during movement [38]. Taken together—

strength, bundle-specific function, and proprioception—the ligament is central to maintaining knee stability, particularly in high-demand athletic tasks.

7.2.2 Stages of anterior cruciate ligament injury

ACL injuries are classified into three primary stages: sprain, partial tear, and complete rupture [245].

1. *Sprain*: A sprain involves overstretching of the ACL fibers without tearing, typically resulting in minimal functional impairment. This type of injury may cause mild pain and swelling but usually does not lead to instability. Athletes with a sprained ACL may still participate in sports but should avoid intense activities until recovery [298].
2. *Partial tear*: A partial tear occurs when some of the ligament fibers are torn, reducing stability and causing pain, particularly during activities that require rapid directional changes. Treatment may involve physical therapy, but surgical intervention may be necessary for individuals seeking full return to sports [245].
3. *Complete rupture*: A complete rupture is a full-thickness tear, leading to significant joint instability. This injury often requires surgical reconstruction, especially for athletes or active individuals. Conservative treatments, such as physical therapy, may not provide sufficient stability for high-demand activities [196].

ACL injuries are most frequently caused by noncontact forces, including abrupt stops, pivots, or directional changes that place extreme stress on the ligament. Proper classification of the injury stage is crucial, as it guides the treatment plan and can affect long-term knee health [196].

7.2.3 Characteristics of magnetic resonance imaging (MRI) modalities

Magnetic resonance imaging (MRI) is the modality of choice for assessing ACL injury because it depicts soft tissues with high contrast, while also rendering adjacent osseous structures [238]. Direct visualization of the ligament helps identify typical injury signs—signal alteration, fiber discontinuity, and periligamentous edema.

1. *T1-weighted imaging*: T1-weighted sequences outline anatomy clearly and are useful for judging ACL morphology and overall integrity, aiding confirmation of structural disruption when present [335].
2. *T2-weighted imaging*: T2-weighted images are sensitive to fluid, making them well-suited to acute settings. Bone bruising and soft-tissue edema that accompany ACL tears are more conspicuous on T2 and can provide indirect evidence of injury [238].
3. *Proton density (PD) imaging*: Proton density-weighted imaging balances soft-tissue contrast and spatial detail. It is particularly helpful for distinguishing partial from complete tears by depicting fiber continuity at high resolution [234].

Advances such as 3D acquisitions and high-field systems (3 Tesla) have improved diagnostic performance for ACL pathology, enabling multiplanar reconstructions and sharper depiction of subtle, partial-thickness injuries [234]. Limitations remain—cost, access, and variability among scanners and protocols can influence consistency—but MRI continues to be the noninvasive reference standard for ACL evaluation because of its detailed soft-tissue visualization [245].

7.3 Methods for anterior cruciate ligament injury detection

7.3.1 Semi-automated methods for anterior cruciate ligament injury detection

Semi-automated approaches combine clinician oversight with computer assistance to streamline ACL assessment. Typical pipelines mark regions of interest, extract descriptive features, and propose candidate injury sites, while the radiologist adjudicates the final call. Published systems vary in how much interaction they require and are commonly compared using sensitivity, specificity, and the degree of user input. In what follows, we outline representative methods and summarize their performance in these terms.

7.3.1.1 Edge detection and thresholding techniques

Early semi-automated work relied on basic image cues—intensity and boundaries—to localize the ligament. Thresholding separates the ACL from adjacent tissues by exploiting signal differences, enabling readers to inspect for fiber discontinuity and other tear indicators [238]. Edge-based operators then highlight putative margins to aid visual tracing. In practice, these procedures are sensitive to noise and acquisition variability and therefore still depend heavily on expert interpretation, particularly when image quality is limited or tears are partial.

7.3.1.2 Region-based segmentation

Region-based segmentation methods are commonly used to delineate anatomical structures within MRI images. One such approach involves initiating from a predefined seed point located within the ACL region and progressively expanding the segmented area based on similar intensity values. The expansion continues until boundaries with differing intensities are encountered, effectively isolating the ligament structure [234]. This technique reduces the impact of noise but requires manual selection of seed points, introducing variability in the segmentation quality. Region-based methods are advantageous because they provide clearer structure boundaries, but they can still struggle with partial tears, where the intensity differences are subtle.

7.3.1.3 Feature extraction and machine learning

Recent advancements have combined feature extraction strategies with machine learning to develop semi-automated diagnostic methods that enhance both robustness and performance in ACL injury detection from MRI scans. In one well-validated study, Histogram of Oriented Gradients (HOG) and GIST descriptors were extracted from manually selected regions encompassing the ACL. These features—capturing ligament structure and textural variations—were then fed into machine learning classifiers like Support Vector Machines (SVM) or Random Forests to distinguish between noninjured, partially torn, and completely ruptured ACLs. The linear-kernel SVM based on HOG achieved an AUC of approximately 0.89 for injury detection and 0.94 for complete ruptures, effectively narrowing down regions of interest for expert review and reducing diagnostic workload. However, this approach remains sensitive to variability across patient anatomy and MRI-acquisition settings [360].

7.3.1.4 Active contour models

Active contour models—commonly referred to as “snakes”–are also employed in semi-automated methods for segmenting the anterior cruciate ligament (ACL). These models minimize energy functions to evolve smooth contours around ligament boundaries, adapting to variations in shape and image texture. While active contour techniques offer flexible and partially automated segmentation, their performance depends heavily on initialization and parameter settings. As a result, performance can deteriorate across differing patient anatomies, and optimization may fail to converge when the ligament margin is poorly contrasted or ill-defined [100].

7.3.1.5 Hybrid methods combining segmentation and feature-based classification

Hybrid pipelines pair a segmentation stage with a learned classifier to gain robustness. In a common setup, the ACL is first delineated using a region-based method; the segmented volume or patch is then passed to a convolutional neural network (CNN) that labels the ligament as intact or injured. This design exploits the precision of (semi-) manual or algorithmic segmentation, while leveraging the discriminative power of CNNs. The trade-offs are practical: Such systems can be computationally intensive and often require advanced expertise to configure and maintain the segmentation and classification steps across sites.

- *Edge detection and thresholding*: Reported sensitivity is roughly 60–70 % and specificity 65–75 %; performance is susceptible to noise, so radiologist input remains high [238].
- *Region-based segmentation*: Typical results show sensitivity of 70–80 % and specificity of 75–85 %. Reader involvement is moderate. Noise is handled better than with simple thresholds, but subtle or partial tears can still be missed [234].

– *Feature extraction + machine learning*: With a reliable feature pipeline, both sensitivity and specificity are generally higher while radiologist effort is reduced; accuracy depends strongly on consistent, high-quality feature generation.

7.3.2 Fully automated methods for anterior cruciate ligament injury detection

Fully automated methods for anterior cruciate ligament (ACL) injury detection utilize deep learning algorithms to interpret magnetic resonance imaging (MRI) scans with minimal human intervention. These systems are particularly advantageous in high-throughput clinical environments where reducing diagnostic workload and turnaround time is critical. Convolutional neural networks (CNNs) have demonstrated strong performance in identifying complex imaging features related to ACL tears, achieving near-radiologist-level accuracy in some studies. By automating the diagnostic process, such models aim to improve efficiency and consistency in clinical decision-making [257].

7.3.2.1 Convolutional neural networks (CNNs)

Convolutional Neural Networks (CNNs) have become the predominant architecture for fully automated anterior cruciate ligament (ACL) tear detection, thanks to their capability to learn structural and spatial features from medical imaging data. A key study employing a CNN trained on MRI scans from the MRNet dataset demonstrated strong diagnostic performance—achieving approximately 96.6 % accuracy, 96.7 % recall, and 95.8 % F1 score in detecting ACL tears [185]. These networks employ successive convolutional layers to automatically extract relevant features—such as ligament discontinuity, texture variations, and anatomical details—without the need for manual feature engineering. One practical strength of CNNs is that they scale: Architectures can be tuned for different in-plane resolutions and extended to volumetric inputs for 3D MRI. The trade-off is that effective training typically needs substantial labeled data, and performance may vary with differences in MRI protocols and image quality across institutions.

7.3.2.2 Transfer learning

Transfer learning is widely used when annotated knee MRI datasets are limited. Models pre-trained on large image repositories (e. g., ImageNet) are adapted by fine-tuning for ACL-focused analysis. In practice, this strategy has improved diagnostic accuracy and strengthened robustness when models are applied to data from different sites, helping them generalize across scanner settings and patient cohorts [101]. By reusing features learned from broad visual domains, transfer learning reduces reliance on large, task-specific ACL datasets and curbs overfitting—benefits that are particularly valuable in routine clinical environments where high-quality labels are scarce.

7.3.2.3 3D convolutional neural networks (3D-CNNs)

In a head-to-head comparison on more than 1,200 MRI exams of knees, overall accuracy was similar for 3D and 2D CNNs (89 % vs 92 %), but the 3D model showed higher sensitivity for partial ACL tears, highlighting the value of volumetric context in clinical diagnostics [191]. Incorporating full 3D-spatial information helps distinguish intact from injured fibers and can reduce missed subtle tears. The downside is resource demand: 3D-CNNs typically require greater compute and memory, which can limit feasibility in some settings.

7.3.2.4 Recurrent neural networks (RNNs) for sequential MRI analysis

Although less common than CNNs, recurrent models offer a way to analyze ordered image series. By modeling dependencies across slice stacks or longitudinal studies, RNNs can capture temporal progression or healing patterns and have reported gains in sensitivity for subtle or diffuse injury presentations. In practice, RNNs are often paired with CNN encoders that extract per-slice features before temporal aggregation. The main constraint is data: Carefully labeled sequential MRI is harder to obtain and annotate at scale.

7.3.2.5 Explainable AI (XAI) in fully automated models

As fully automated systems gain accuracy, interpretability remains a gating factor for clinical adoption. High-performing CNNs can behave like "black boxes," which undermines confidence among readers who must justify decisions. Explainable AI tools—such as saliency maps, Gradient-weighted Class Activation Mapping (Grad-CAM), and attention-based visualizations—aim to reveal the image regions most responsible for a prediction, improving transparency and clinician trust. Clear, clinically relevant explanations are essential for routine use, so these methods should be validated with expert input and embedded in workflows to support reliable decision-making.

Explainable artificial intelligence (XAI) techniques have been incorporated into CNN-based models for ACL injury detection, enabling radiologists to see which image regions influence the model's decisions. For example, Grad-CAM generates heat maps that highlight areas of diagnostic importance, thus improving transparency and building clinician trust in automated systems [246]. Although XAI improves the interpretability of deep learning models in medical imaging, further development of customized methods for ACL assessment is needed to ensure that the highlighted regions are clinically relevant and useful.

- *Convolutional neural networks (CNNs)*: By learning hierarchical image features from MRI, CNNs can detect ACL tears with high accuracy and can be configured for different in-plane resolutions. The main caveats are the need for sizeable labeled cohorts and sensitivity to protocol and quality differences across scanners [185].
- *Transfer learning*: Reusing weights from large, pretrained models helps systems generalize on ACL-focused tasks when labeled data are limited. In clinical settings,

this approach shortens training, reduces overfitting risk, and lowers the barrier to deployment when annotations are scarce [101].

- *3D convolutional neural networks (3D-CNNs)*: Operating on volumetric MRI, 3D-CNNs integrate information across slices and are effective for subtle tear patterns. Their drawback is resource demand—greater memory and compute can constrain use in some clinics [191].
- *Recurrent neural networks (RNNs)*: Modeling slice sequences or longitudinal studies, RNNs can improve sensitivity by capturing temporal relationships. The approach, however, depends on well-labeled sequential datasets, which are harder to curate at scale.
- *Explainable AI (XAI)*: Techniques such as saliency maps and Grad-CAM increase interpretability by highlighting image regions that drive predictions, enabling radiologists to verify model reasoning and supporting clinical acceptance.

Fully automated pipelines for ACL injury detection now approach radiologist-level performance in several reports. Remaining challenges—limited labeled data, computational overhead, and the need for transparent decision support—continue to shape the field. Ongoing work is likely to emphasize improved data access and curation, efficiency on routine hardware, and stronger explainability to make end-to-end ACL detection practical in day-to-day diagnostics.

7.3.3 Conclusion

Semi-automated methods for ACL injury detection are valuable tools for assisting radiologists, each with unique strengths and limitations. Simple techniques like edge detection require significant manual input and may suffer from noise, while advanced approaches such as hybrid segmentation-classification models offer high accuracy with minimal involvement but require computational resources. Fully automated methods, particularly those based on deep learning, have advanced the field by enabling faster, high-accuracy diagnostics without the need for extensive radiologist involvement. Techniques such as CNNs, 3D-CNNs, and RNNs offer promising results in accurately identifying ACL injuries, while transfer learning and explainable AI (XAI) approaches enhance model generalizability and interpretability, respectively.

Despite their effectiveness, fully automated methods face challenges in interpretability, data requirements, and computational demands. These models require large, well-annotated datasets, which can be difficult to obtain in clinical settings, and their computational needs may limit their deployment in some healthcare environments. Although explainable AI improves transparency, fully automated systems still must earn clinician confidence by offering clearer, clinically meaningful rationales for their predictions.

Looking ahead, a practical direction is to unify semi-automated and fully automated pipelines so they complement one another. Such a framework would retain the interpretability and explicit clinician oversight of semi-automated methods, while harnessing the speed and consistency of end-to-end models. Parallel efforts should target robustness—designing models that adapt to heterogeneous MRI data from different scanners and sites—so that generalization and clinical relevance improve. In sum, a balanced design that jointly optimizes accuracy, interpretability, and computational efficiency is likely to move ACL injury diagnostics toward routine, reliable use and better patient outcomes in orthopedic care.

7.4 Performance evaluation measures

Performance evaluation for detecting and classifying anterior cruciate ligament (ACL) injuries relies on various quantitative metrics to assess model effectiveness. Common measures include accuracy, sensitivity, specificity, precision, F1 score, and the area under the receiver operating characteristic curve (AUC) [191]. Additionally, metrics like the Matthews Correlation Coefficient (MCC) and Cohen's Kappa have been adopted to offer deeper insights, especially when dealing with imbalanced datasets.

1. *Accuracy*: The accuracy metric measures a model's overall performance by calculating the proportion of correctly classified cases, including both true positives and true negatives, relative to all evaluated cases. While high accuracy is generally preferred, it can be deceptive in imbalanced datasets, where a model may accurately classify the majority class but struggle with the minority class.
2. *Sensitivity (recall)*: Sensitivity, also known as the true positive rate, reflects the model's ability to accurately detect positive cases among all actual positives. In medical contexts, this metric is crucial, as a low sensitivity could result in missed diagnoses of ACL tears, potentially delaying necessary treatment. High sensitivity is often prioritized in models where it is essential to minimize false negatives [251].
3. *Specificity*: Specificity refers to the model's ability to correctly identify true negatives. This is important to avoid overdiagnosing ACL injuries and helps in minimizing false positives, especially in fully automated methods that may flag noninjured ACLs as damaged.
4. *Precision*: Also known as positive predictive value, precision is the proportion of predicted positives that are truly positive. High precision means few false alarms, which helps avoid unnecessary follow-ups or treatment in clinical care.
5. *F1 score*: The harmonic mean of precision and recall. It provides a single summary that balances missed detections and false positives and is especially informative when classes are imbalanced. A high F1 indicates strong detection while keeping incorrect positives low.
6. *AUC*: The area under the receiver-operating characteristic curve summarizes class separability across all thresholds by relating true-positive to false-positive rates. Larger AUC indicates better discrimination and is widely used for diagnostic modeling, including distinguishing ACL injury from normal cases [191].

7. *Matthews correlation coefficient (MCC)*: A correlation-style metric computed from all four entries of the confusion matrix (TP, TN, FP, FN). Unlike accuracy or precision alone, MCC remains informative with class imbalance and offers a balanced view of prediction quality [52].

These metrics enable researchers and clinicians to comprehensively assess model performance and identify potential limitations in diagnostic accuracy. Each performance measure provides specific insights and plays a role depending on the clinical context and application goals. In scenarios where patient safety is paramount, such as ACL injury detection, sensitivity often takes precedence, as missing an injury could lead to deteriorating knee function or chronic issues. Specificity, on the other hand, is important for reducing the risk of unnecessary procedures or treatments that may arise from false positives, making it particularly useful in semi-automated settings where human oversight is available.

The F1 score and the area under the receiver operating characteristic curve (AUC) are key metrics for evaluating and selecting machine learning models in healthcare. The F1 score balances precision and sensitivity, making it useful when false positives and false negatives have distinct clinical consequences. The AUC assesses a model's ability to differentiate between injury and noninjury cases across various thresholds. Together, these metrics help clinicians select models that are both balanced and effective at distinguishing cases, ensuring reliable use in medical practice [186]. Metrics like MCC and Cohen's Kappa are particularly valuable in cases with high class imbalance since they provide additional insight into model reliability beyond simple accuracy [52].

7.5 Comparative analysis of the methods performance

7.5.1 Comparative analysis of quantitative results

Quantitative comparisons between semi-automated and fully automated systems for ACL injury detection point to clear trade-offs in both performance and usability. End-to-end deep models—most often CNN-based—have reported high sensitivity and specificity, with several studies noting diagnostic accuracies above 90 % on validated MRI datasets [253]. Their appeal is strongest in high-throughput settings, where rapid and consistent reads are at a premium.

Semi-automated workflows, by contrast, benefit from direct clinician oversight. Expert input can trim false positives and negatives in difficult cases, e. g., when image quality is suboptimal or findings are atypical, but this comes with added interaction time and variability tied to reader experience. In aggregate, fully automated methods typically lead on speed and scalability, whereas semi-automated approaches may inspire greater trust when interpretability and case-by-case adjudication are critical.

7.5.2 Comparative analysis of qualitative results

From a qualitative standpoint, interpretability and usability of diagnostic systems are essential for clinical adoption. Semi-automated methods often provide clinicians with greater control over the analysis, which is particularly valuable when interpreting subtle differences in MRI scans that may not be captured by algorithms. By contrast, fully automated systems benefit from the integration of Explainable AI (XAI) techniques—such as gradient-based attention maps—that visually indicate which regions of an image contributed most to the model's decision. These visual explanations enhance transparency and help clinicians validate the AI system's reasoning, fostering greater trust in automated diagnostic tools [246]. However, fully automated methods face challenges in interpretability, as complex models like deep neural networks can operate as "black boxes," where the underlying decision process is not immediately clear. XAI methods attempt to bridge this gap, but there remains a need for more transparent models, especially in clinical environments where diagnostic decisions can have significant consequences.

7.5.3 Analysis of the effectiveness and limitations

Both semi-automated and fully automated methods have distinct effectiveness and limitations. Semi-automated methods, while labor-intensive, allow for detailed, nuanced interpretation, making them valuable in complex cases or in settings where high diagnostic precision is required. However, the reliance on human input may introduce variability and slow down the diagnostic process, making it less feasible for high-throughput clinical environments.

Fully automated methods, on the other hand, offer scalability, speed, and reduced dependency on human operators, making them well-suited for hospitals with large patient volumes or limited radiologist availability. Nonetheless, end-to-end models depend on large, high-quality training datasets and can falter on atypical presentations or outliers [327]. Beyond headline metrics, their "black-box" character and occasional misses on subtle findings strengthen the case for hybrid designs that combine clinician expertise with AI, thereby improving diagnostic robustness.

7.6 Case studies

7.6.1 Case studies illustrating challenges for automated anterior cruciate ligament injury detection

Automated ACL detection systems show strong promise, yet deployment in routine practice exposes several stress points. Case studies repeatedly surface issues tied to image quality, heterogeneity in MRI protocols, and patient-specific anatomy. These reports are useful not only as cautionary tales but also as probes of how robust and adaptable current models are under real-world variation.

A representative multicenter analysis trained a convolutional network on data from a single site and then evaluated it on external cohorts acquired with different scanners and protocols. Sensitivity and specificity dropped markedly on the outside datasets, underscoring the impact of domain shift and the need for training on heterogeneous, multicenter collections to improve generalization [227]. In practice, careful dataset curation, protocol harmonization, and explicit cross-site validation are essential safeguards.

Interpretability is equally important for clinical uptake: Readers must be able to see why a model reached a conclusion and judge whether the highlighted evidence is clinically convincing. Although convolutional neural networks (CNNs) often deliver high diagnostic performance, their internal decision-making is typically opaque, which can undermine clinician confidence. This "black-box" nature poses a barrier to integration in routine clinical workflows. To address this, explainable AI (XAI) techniques–such as saliency mapping and concept activation methods—have been introduced to visually highlight regions in medical images that support model predictions. Rigorous validation of such explanations in collaboration with healthcare professionals is essential to bridge the gap between model transparency and clinical relevance [308].

7.6.2 Lessons learned from these cases

The case studies above provide crucial lessons for developing and deploying automated ACL injury detection systems:

1. *Diversity in training data*: The importance of diverse training datasets cannot be overstated. Models trained on single-institution data may lack robustness when deployed in varied clinical settings with different MRI machines and protocols. Incorporating data from multiple sources improves the generalizability of models and reduces potential biases.
2. *Image quality and preprocessing requirements*: Automated systems are highly sensitive to image quality, and low-resolution or noisy images can lead to significant performance drops. This suggests a need for standardized preprocessing techniques to normalize image quality and possibly pretrain models on lower-quality images to increase resilience [251].
3. *Complex injury types*: Detecting partial ACL tears and degenerative changes remains a challenge since these injuries often lack clear morphological markers. Integrating multiview MRI analysis, or utilizing advanced architectures like 3D CNNs, can help improve sensitivity for these complex injury types.
4. *Importance of explainable AI (XAI)*: Explainability is vital in clinical applications since it enables clinicians to trust AI systems by understanding their decision-making processes. However, as demonstrated in [101], explainable models need

refinement to ensure that highlighted regions are clinically relevant. Collaboration with clinicians to validate XAI outputs can enhance model reliability and trust.
5. *Continuous model adaptation and feedback loops*: Clinical deployment of AI models is not a one-time event; models must be adapted based on real-world feedback. Regular updates to account for new data, imaging techniques, and clinician input are essential for maintaining accuracy and relevance.

Taken together, these findings make clear that building automated ACL detection that holds up in practice is nontrivial. Progress will depend on tight clinician–engineer collaboration, robust multisite data curation, and algorithms designed to adapt to the variability of real clinical environments.

7.7 Future directions

The field of automated ACL injury detection has advanced quickly, yet several gaps remain before routine clinical use is feasible. This section outlines practical directions for future work, with emphasis on stronger cross-site generalization, integration of multimodal information, clearer model interpretability, and personalization strategies that tailor predictions to patient-specific factors.

7.7.1 Improving model generalization across diverse clinical settings

A central deployment challenge is robustness across institutions. Models that perform well at the development site often face protocol and hardware shifts elsewhere, leading to performance drift. Improving generalization calls for training on large, heterogeneous datasets drawn from multiple scanners and centers, coupled with explicit cross-site validation. Techniques such as domain adaptation and transfer learning can help align feature representations across acquisition settings. Future work should prioritize pipelines that are resilient to scanner variability and protocol differences so that accuracy remains stable across clinical environments and, in turn, clinician confidence increases.

7.7.2 Integration of multimodal and longitudinal data

Combining different types of imaging data—such as T1-weighted, T2-weighted, and proton density-weighted MRI—with clinical history or longitudinal studies can significantly improve the accuracy and robustness of ACL detection models. Multimodal fusion approaches have been shown to enhance sensitivity, particularly for identifying subtle ligament injuries or early degenerative changes that single-modality models often miss.

For instance, a recent radiomics study demonstrated that a support vector machine classifier trained on both T1 and proton density MRI achieved AUC scores above 0.97 and sensitivity over 0.93 for ACL tear detection, outperforming single-sequence models [51].

7.7.3 Enhancing model interpretability and explainable AI (XAI) techniques

Interpretability of the model is essential to foster trust among clinicians and facilitate safe integration of artificial intelligence (AI) systems into clinical workflows. Explainable artificial intelligence (XAI) methods, such as saliency maps and feature attribution techniques, allow models to highlight key areas in MRI scans relevant to diagnosis, making results clearer and more practical for clinicians. Future work should center on XAI tools that are clinically meaningful and codesigned with healthcare professionals, ensuring alignment with medical reasoning and routine workflows. Improving the explainability of models for anterior cruciate ligament (ACL) injury detection, in particular, can strengthen clinician trust and accelerate the adoption of AI in diagnostic decision-making.

7.7.4 Developing personalized diagnostic models

Personalized diagnostics—models tuned to a patient's anatomy and risk profile—are a natural next step for medical AI. For ACL assessment, future work could develop predictors that adapt to individual characteristics such as age, activity level, and ligament morphology, rather than assuming a one-size-fits-all template. By explicitly modeling these factors, systems can deliver more accurate, patient-specific injury assessments and tailored treatment recommendations, which should translate into better outcomes. Embedding this paradigm within automated ACL detection would mark a shift from generalized to individualized decision support, increasing clinical utility.

7.7.5 Leveraging federated learning for data privacy and access

Federated learning enables institutions to train shared models without pooling protected patient data, addressing privacy and governance constraints that are acute in medical imaging. By coordinating learning across heterogeneous cohorts—with differences in population, scanners, and acquisition protocols—this decentralized setup can improve generalization and clinical robustness for diagnostic systems, including those aimed at anterior cruciate ligament (ACL) injury detection. Moving forward, there is a need for FL frameworks tailored to MRI-based musculoskeletal applications, with attention to secure aggregation, protocol harmonization, and rigorous cross-site validation [126].

7.7.6 Summary of key points

The future of automated ACL injury detection systems will likely be shaped by the following key advancements:

1. *Model generalization*: Developing models that can generalize across diverse clinical settings and imaging protocols will be essential for reliable, widespread deployment.
2. *Multimodal and longitudinal data integration*: Incorporating various imaging modalities and longitudinal data can improve the accuracy of detecting subtle injuries and tracking recovery progress.
3. *Explainable AI (XAI)*: Advancing XAI techniques to make model outputs clinically interpretable will help increase clinician trust and integration into medical decision-making.
4. *Personalized diagnostics*: Tailoring models to individual patient characteristics could improve diagnostic accuracy and enhance the clinical utility of automated ACL injury detection.
5. *Federated learning for data privacy*: Utilizing federated learning can expand dataset diversity while preserving data privacy, addressing key challenges in medical AI research [327].

These directions represent essential steps toward refining automated ACL injury detection systems, ensuring they are clinically valuable, widely generalizable, and ethically robust. The integration of these advancements could make AI-based diagnostics a transformative tool in orthopedic imaging and sports medicine.

8 Computer-aided diagnosis of breast-cancer detection: challenges and solutions

Abstract: Breast cancer remains one of the leading causes of mortality among women worldwide, and early detection is strongly associated with improved survival. Medical imaging has become central to screening and diagnosis, yet manual interpretation is labor-intensive and prone to variability. Recent advances in artificial intelligence have enabled computer-aided systems that support radiologists by analysing mammography, ultrasound, MRI, and histopathology images. This chapter reviews the evolution from classical machine-learning pipelines to deep convolutional and transformer-based architectures, highlighting representative studies, datasets, and evaluation metrics. Special attention is given to hybrid models that combine complementary features for improved accuracy and robustness. In addition, the chapter examines practical challenges—such as data imbalance, generalization across scanners, and clinical integration—and outlines promising directions including lightweight deployment, explainability, and multimodal fusion. Together, these insights provide a consolidated view of how AI can accelerate progress toward reliable and accessible breast cancer detection.

Keywords: Breast cancer imaging, Computer-aided diagnosis (CAD), Deep learning, Transformer architectures, Biomedical image analysis

8.1 Introduction

Breast cancer is one of the most common cancers worldwide and remains a major cause of cancer-related mortality, with early and accurate detection directly linked to better survival rates [13, 281]. Medical imaging plays a central role in both screening and diagnosis, with modalities such as mammography, ultrasound, magnetic resonance imaging (MRI), and histopathology being routinely used in clinical workflows [74, 83].

Despite their clinical utility, manual interpretation of medical images is time-consuming and subject to inter-observer variability, which may lead to diagnostic inconsistencies [92, 348]. This has motivated the development of computer-aided diagnosis (CAD) systems to assist radiologists in image interpretation and decision-making. Early CAD approaches often relied on handcrafted feature extraction combined with traditional machine-learning algorithms [34, 348], but these methods were limited in their ability to capture complex patterns in breast cancer images.

Over the past decade, deep learning has transformed breast cancer image analysis. Convolutional neural networks (CNNs) demonstrated strong performance by enabling end-to-end feature learning from raw data [153, 232, 355]. More recently, transformer-based architectures have been introduced to model long-range dependencies and con-

https://doi.org/10.1515/9783111389059-008

textual relationships in imaging data [70, 161, 316], while hybrid CNN–transformer frameworks have emerged to combine local feature extraction with global attention mechanisms for improved accuracy [48, 82, 99].

These advances have produced state-of-the-art results across modalities including mammography, ultrasound, MRI, and histopathology [53, 184, 314]. Nonetheless, significant challenges remain in terms of limited annotated datasets, cross-domain generalization, model interpretability, and deployment in real-world clinical environments. This chapter provides a comprehensive review of AI-based approaches for breast cancer detection and diagnosis, organized around CNNs, transformer architectures, and hybrid models. It also highlights widely used datasets, evaluation metrics, comparative analyses, challenges, and future research directions.

8.2 Imaging modalities and datasets

Several imaging modalities are routinely used for breast-cancer detection and diagnosis, each with distinct strengths and limitations. These modalities not only drive clinical decision-making but also serve as the foundation for developing and validating AI-based computer-aided diagnosis (CAD) systems [83, 184].

8.2.1 Mammography

Mammography is the most widely adopted screening tool worldwide and remains the clinical standard for early breast cancer detection. It provides high-resolution, low-dose X-ray images that are effective in identifying microcalcifications and subtle structural changes [281]. However, its sensitivity decreases in patients with dense breast tissue, leading to potential false negatives.

Public datasets such as the Digital Database for Screening Mammography (DDSM) and INbreast have been widely used in AI research, providing benchmarks for algorithm development [83]. In addition, curated datasets for breast density prediction and lesion detection, such as those used by Lee et al. [146], have supported more advanced deep learning studies.

8.2.2 Ultrasound

Breast ultrasound is commonly used as a complementary modality, especially in women with dense breast tissue where mammography is less reliable. It is non-invasive and radiation-free but highly operator-dependent and prone to variability [328].

The BUSI dataset (Breast Ultrasound Images) includes 780 images labeled into normal, benign, and malignant categories, making it a widely used benchmark for segmentation and classification tasks. Recent deep learning approaches, such as HCTNet [99], have used BUSI and related ultrasound datasets to advance breast lesion analysis.

8.2.3 Magnetic resonance imaging (MRI)

MRI provides high sensitivity and detailed functional and structural information about breast tissue. It is recommended for high-risk patients and in preoperative staging [74]. Despite its strengths, MRI is expensive, time-consuming, and less accessible in low-resource settings.

Publicly available datasets hosted in The Cancer Imaging Archive (TCIA) include breast MRI collections that have been employed in AI-based breast cancer studies [184]. These datasets make possible the evaluation of models for tumor detection, segmentation, and staging.

8.2.4 Histopathology

Histopathology remains the gold standard for definitive breast-cancer diagnosis, offering cellular-level details for tumor grading and classification [74].

- *BreakHis:* The BreaKHis dataset contains 7,909 images from 82 patients across four magnification levels (40×, 100×, 200×, 400×), annotated as benign or malignant. It is one of the most widely used benchmarks in deep learning-based histopathological breast-cancer research [45, 270].
- *BACH:* The Breast Cancer Histology dataset was introduced as part of the BACH grand challenge. It consists of hematoxylin and eosin (H&E) stained microscopic images categorized into normal tissue, benign lesions, in-situ carcinoma, and invasive carcinoma [23].

These histopathology datasets have enabled the development of CNNs, transformers, and hybrid models for tumor classification, segmentation, and grading.

8.2.5 Benchmark datasets summary

Table 8.1 summarizes commonly used public datasets across different breast cancer-imaging modalities. These datasets have been crucial for reproducible AI research, model benchmarking, and clinical validation [83, 184].

Table 8.1: Summary of commonly used breast cancer-imaging datasets.

Dataset	Modality	Size (approx.)	Ref.
DDSM	Mammography	2,500 studies	[83]
INbreast	Mammography	115 cases	[83]
BUSI	Ultrasound	780 images	[328]
TCIA collections	MRI	Varies	[184]
BreakHis	Histopathology	7,909 images	[45, 270]
BACH	Histopathology	400 images	[23]

8.3 AI approaches for breast-cancer imaging

Deep learning has become a cornerstone for breast cancer-imaging analysis. Traditional image processing techniques had limited ability to capture the complex structures of lesions, but AI-based methods—especially convolutional and transformer-based models—have demonstrated significant improvements in both accuracy and robustness. This section reviews key approaches, grouped by architecture.

8.3.1 Convolutional neural networks (CNNs)

CNNs are among the most widely used models in breast-cancer detection, particularly for mammography, ultrasound, and histopathological images. Early works applied standard CNNs to biopsy images and demonstrated improved classification performance compared to traditional machine-learning models [270, 348].

Advanced CNN designs such as U-Net and its extensions have also been widely adopted for segmentation of breast lesions. The U-Net architecture was first introduced for biomedical segmentation tasks [232] and later refined in UNet++ to improve multi-scale feature representation [355]. Feature Pyramid Networks (FPNs) [153] and encoder–decoder architectures with atrous convolution [49] further enhanced detection of small lesions in mammographic and ultrasound images.

Histopathology-focused CNNs, including magnification-invariant models [34] and structured deep-learning frameworks [92], provided strong results across datasets such as BreaKHis [45]. These methods underscore the adaptability of CNNs across imaging modalities.

8.3.2 Hybrid and attention-enhanced CNNs

Recent models incorporate residual and recurrent connections into CNNs to boost performance. Inception-based hybrids have been explored for breast cancer-histopathology classification [9]. Attention mechanisms were also integrated into CNNs to refine feature extraction, especially for dense breast tissue where lesion localization is challenging [274].

Mobile and lightweight networks are gaining importance for clinical deployment. Examples include efficient CNN backbones that balance accuracy with computational feasibility, making them suitable for large-scale screening programs.

8.3.3 Transformer-based models

Transformers, originally developed for natural language processing, have rapidly advanced in medical imaging. Vision Transformers (ViT) have been adapted for breast-cancer classification [70], and mammogram classification has been explored using vision-transformer-based transfer learning [21].

Hybrid CNN-transformer models such as TransUNet [48], UTNet [82], and nnFormer [351] combine the strengths of convolutional layers for local feature extraction with transformer modules for global context modeling. These methods outperform standard CNNs in segmentation and classification tasks.

Specialized transformer architectures like Swin Transformer [161] and Pyramid Vision Transformer (PVT) [316] have further improved multiscale feature representation and computational efficiency. Recent ultrasound segmentation studies have adopted CNN-transformer hybrids, such as HCTNet [99], to achieve state-of-the-art performance.

Overall, CNNs remain the foundation of breast cancer-imaging analysis, while transformer-based models and hybrid architectures represent the next frontier. CNNs excel in localized feature extraction, whereas transformers enhance global contextual reasoning. Together, these approaches form the basis for high-performance diagnostic tools that are increasingly practical for clinical use.

8.4 Case studies on AI approaches

This section highlights representative works applying deep learning and hybrid architectures to breast-cancer imaging. Each case study summarizes the dataset (as described in Section 8.1), methodology, and outcomes, providing insights into the strengths and limitations of current approaches.

8.4.1 Histopathology classification

Spanhol et al. [270] applied CNNs to the BreaKHis dataset, which contains 7,909 images across four magnification levels. Their CNN-based approach achieved competitive accuracy in distinguishing benign and malignant lesions, demonstrating the feasibility of deep learning for histopathology. However, performance varied across magnification levels, and generalization beyond the dataset remained a challenge. Subsequent studies have attempted to improve robustness using magnification-independent strategies and ensemble classifiers [9, 34].

8.4.2 Multi-label transfer learning

Chougrad et al. [53] introduced a multi-label transfer learning framework for early breast cancer diagnosis. Their model leveraged pretrained CNNs and was validated on mammographic datasets, achieving improved sensitivity and specificity compared to traditional approaches. A key strength of this method is its robustness to varying imaging conditions, although the requirement of large-scale annotated data for fine-tuning remains a limitation.

8.4.3 Hybrid CNN–Transformer segmentation

Chen et al. [48] presented TransUNet, a hybrid model that integrates convolutional encoders with Transformer-based self-attention mechanisms. Applied to medical image segmentation tasks, including breast-lesion analysis, TransUNet outperformed conventional U-Net architectures by capturing both local spatial features and long-range dependencies. Gao et al. [82] proposed a related UTNet architecture, and He et al. [99] designed HCTNet for ultrasound segmentation. These methods demonstrated higher Dice scores and improved lesion-boundary delineation, though at the cost of increased computational complexity.

8.4.4 Transformer and detector-based approaches for mammograms

Su et al. [274] developed YOLO-LOGO, a Transformer-based YOLO segmentation model tailored for breast mass detection in mammograms. The framework achieved superior detection accuracy compared to CNN-only models, particularly in localizing small lesions. Similarly, Ayana et al. [21] employed Vision Transformers (ViT) for mammogram classification, demonstrating significant improvements in generalization. While transformer models capture rich contextual features, they typically require larger training datasets and higher computational resources than CNN-based approaches.

8.4.5 Summary of case studies

Table 8.2 provides a concise comparison of these representative case studies, including datasets (see Sect. 8.1), methods, and outcomes.

8.5 Comparative analysis

To evaluate the effectiveness of multiple AI approaches for breast-cancer detection and classification, it is important to compare results across diverse studies. The comparative insights presented in this section are compiled from the published literature and not from new experiments conducted by the authors. Since results are reported on various datasets, they should be interpreted as indicative rather than strict head-to-head benchmarks (see Sect. 8.1 for dataset details).

Table 8.2: Summary of representative case studies on AI-based breast-cancer imaging.

Author(s)	Dataset (see Sec. 8.1)	Method	Key Results / Observations
Spanhol et al. [270]	BreaKHis	CNN-based classification	Achieved high accuracy across magnification levels; generalization to external datasets remained limited.
Chougrad et al. [53]	Mammography datasets	Multi-label transfer learning using pretrained CNNs	Improved sensitivity and specificity; robust to imaging variations but required large annotated data.
Chen et al. [48], Gao et al. [82], He et al. [99]	MRI, ultrasound, histopathology	Hybrid CNN–Transformer segmentation (TransUNet, UTNet, HCTNet)	Outperformed U-Net with higher Dice scores and better lesion-boundary delineation; increased computational demands.
Su et al. [274], Ayana et al. [21]	Mammograms	Transformer-based detectors (YOLO-LOGO, ViT models)	Superior detection accuracy and generalization; effective in small lesion localization but required more data and computational resources.

8.5.1 Performance across architectures

Deep learning models vary significantly in predictive accuracy, computational cost, and parameter requirements. Figure 8.1 presents a comparison of representative CNNs, hybrid networks, and Transformer-based architectures, based on results reported in prior studies. The heatmap illustrates the trade-off between accuracy and computational com-

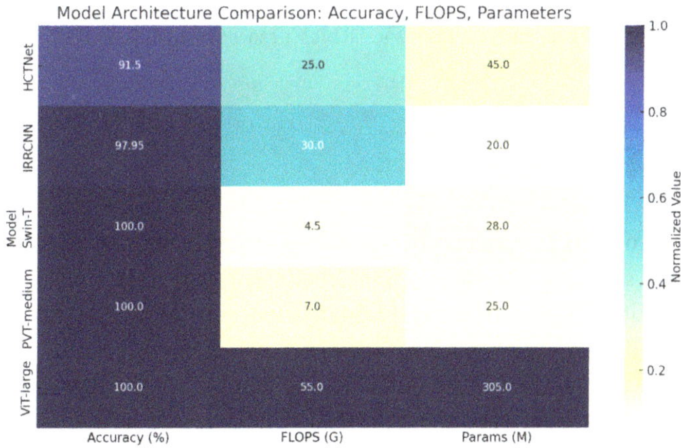

Figure 8.1: Comparison of accuracy, FLOPs, and parameter counts across representative models, compiled from the published literature. Results are reported on different datasets and should be interpreted as indicative performance trends, not direct benchmarks.

plexity: While Transformer-based models and hybrid CNN–Transformer frameworks often achieve superior accuracy, they require substantially more FLOPs and parameters, increasing the computational cost [9, 21, 99, 161].

8.5.2 Effect of data augmentation and preprocessing

Preprocessing and data-augmentation strategies play a vital role in enhancing generalization. Techniques such as rotation, flipping, and color normalization are commonly reported to reduce overfitting and improve classification robustness. Figure 8.2 summarizes published findings that demonstrate how augmentation improves stability in training, particularly for small or imbalanced datasets [53, 270]. These results underline the importance of preprocessing pipelines in real-world AI deployments.

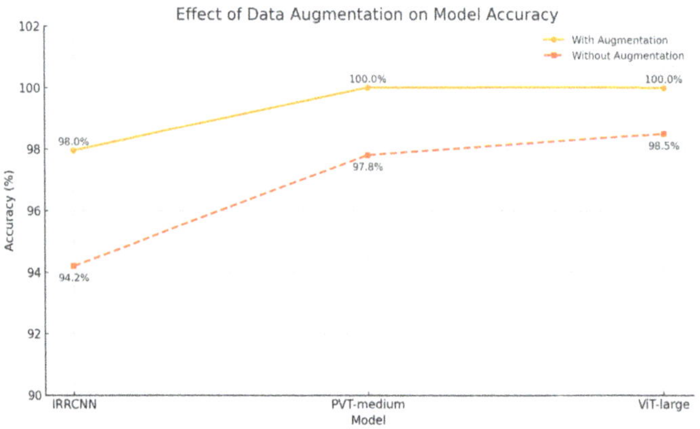

Figure 8.2: Impact of data augmentation on model performance in breast-cancer imaging tasks, compiled from the published results on numerous datasets [53, 270].

8.5.3 Comparative metrics across studies

Reported performance across studies can also be contrasted using standard evaluation metrics such as accuracy, sensitivity, specificity, Dice coefficient, and AUC. Figure 8.3 presents comparative results extracted from representative works, showing that, while CNNs remain strong baselines, hybrid and Transformer-based methods consistently achieve superior Dice and AUC scores. These outcomes, as reported in the literature, suggest that next-generation architectures provide more reliable lesion segmentation and classification [21, 48, 82, 274].

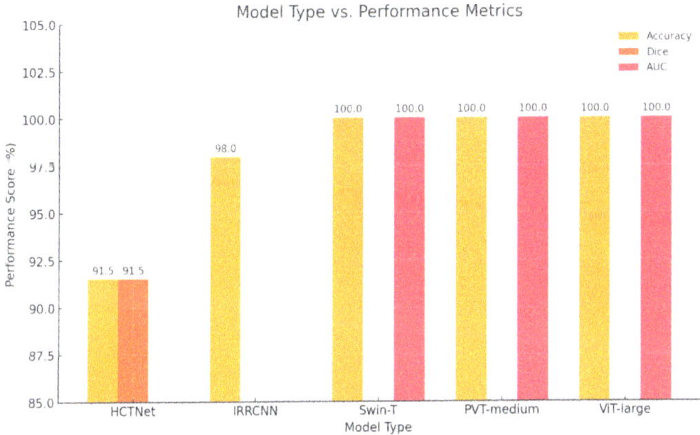

Figure 8.3: Comparative performance of representative AI models on breast-cancer datasets, compiled from the published literature. Results come from numerous datasets and are indicative of general trends rather than direct benchmarks.

8.5.4 Representative studies overview

Table 8.3 maps representative models to the datasets they were evaluated on, contextualizing the results shown in Figs. 8.1–8.3. Dataset details are summarized in Sect. 8.1.

Table 8.3: Representative AI models for breast cancer-imaging with corresponding datasets. Results in Figs. 8.1–8.3 are drawn from these and other published works.

Author(s)	Dataset(s)	Model / Method
Spanhol et al. [270]	BreaKHis (histopathology)	CNN for benign vs malignant classification
Chougrad et al. [53]	Mammography datasets	Multi-label transfer learning (pretrained CNNs)
Chen et al. [48]	MRI, histopathology	TransUNet (CNN + Transformer hybrid)
Gao et al. [82]	MRI, ultrasound	UTNet (CNN + Transformer hybrid)
He et al. [99]	BUSI (ultrasound)	HCTNet (hybrid CNN–Transformer)
Su et al. [274]	Mammograms	YOLO-LOGO (Transformer-based YOLO)
Ayana et al. [21]	Mammograms	Vision Transformer (ViT) for classification
Alom et al. [9]	Histopathology	Inception recurrent residual CNN
Liu et al. [161]	General vision datasets; adapted to breast imaging	Swin Transformer backbone

8.6 Challenges and limitations

Despite remarkable advances in automated breast-cancer detection and diagnosis using deep learning, several challenges remain unresolved. These limitations hinder real-world deployment and affect the reliability of proposed solutions. Figure 8.4 summarizes some of the critical challenges discussed in this section.

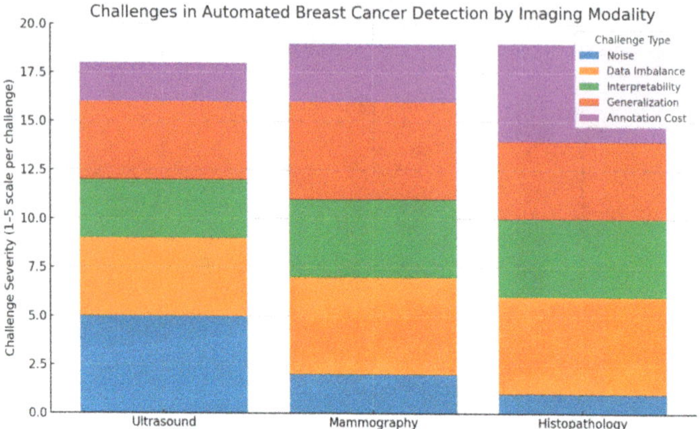

Figure 8.4: Overview of challenges and limitations in automated breast-cancer detection.

8.6.1 Data quality and availability

One of the foremost challenges is the limited availability of large and diverse annotated datasets. While popular datasets such as BreaKHis [270] and BACH [23] have advanced research, they still do not capture the full diversity of patient demographics, imaging modalities, and clinical conditions. Furthermore, annotation quality depends heavily on expert pathologists, making it resource-intensive and prone to interobserver variability.

8.6.2 Model generalizability

Most deep-learning models exhibit strong performance on benchmark datasets but struggle when applied to external data from different institutions. This lack of generalizability often arises due to variations in imaging protocols, devices, and patient populations. Figure 8.5 illustrates performance discrepancies observed when models trained on one dataset are tested on another, highlighting the domain-shift problem.

8.6.3 Interpretability and clinical acceptance

Deep-learning models are often criticized as "black boxes" due to their lack of interpretability. In clinical practice, radiologists and oncologists require transparent decision-making systems that can provide reasoning or highlight evidence for their predictions. Without explainability, it is difficult to gain trust and ensure ethical responsibility in deploying AI-based systems.

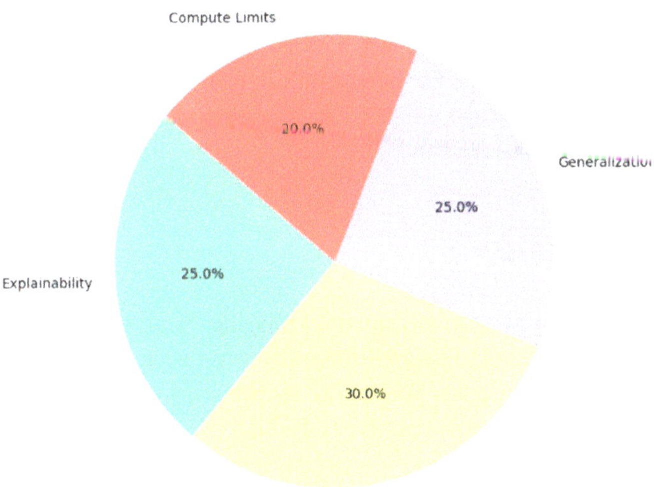

Figure 8.5: Illustration of model generalizability issues across datasets.

8.6.4 Computational and resource constraints

Training advanced architectures such as Vision Transformers (ViT) [70] or hybrid CNN-Transformer models requires significant computational resources, which may not be accessible in many clinical settings. Additionally, high storage and memory demands pose challenges for real-time applications, especially in low-resource healthcare environments.

8.6.5 Ethical and privacy concerns

Breast-cancer datasets involve sensitive medical information, raising issues related to privacy and security. Ensuring patient confidentiality, while promoting open data sharing, remains a delicate balance.

In summary, while AI models for breast-cancer detection have achieved state-of-the-art accuracy, challenges in data diversity, generalizability, interpretability, resource availability, and ethics continue to hinder their translation into real-world practice. Addressing these issues requires interdisciplinary collaboration among computer scientists, clinicians, ethicists, and policymakers.

8.7 Future directions

Although current advances in deep learning have significantly improved breast-cancer detection and diagnosis, there remains substantial scope for future research.

8.7.1 Larger and more diverse datasets

To overcome the limitations of current benchmark datasets, future research must focus on collecting larger, multi-institutional, and demographically diverse datasets. Such datasets should represent variations in age, ethnicity, imaging devices, and disease subtypes, thereby improving model robustness and fairness.

8.7.2 Integration of multimodal data

Breast-cancer diagnosis often relies on a combination of imaging, histopathology, and genomic data. Future AI models should integrate these heterogeneous modalities to provide comprehensive diagnostic insights. Multimodal learning approaches can potentially capture subtle correlations across data types, leading to more accurate and personalized predictions.

8.7.3 Explainable and trustworthy AI

Interpretability remains a critical challenge. Future models must incorporate explainable AI (XAI) techniques that highlight decision-making processes in a manner understandable to clinicians. Visual saliency maps, attention mechanisms, and feature attribution methods can bridge the gap between model predictions and clinical trust [101].

8.7.4 Federated and privacy-preserving learning

Federated learning and differential privacy frameworks enable collaborative model training across institutions without compromising patient confidentiality. These approaches can significantly increase data diversity while adhering to regulatory guidelines such as HIPAA and GDPR [256]. Figure 8.6 shows a federated learning workflow with multiple local clients training models and a central server aggregating updates.

8.7.5 Towards deployment and clinical integration

For AI systems to achieve large-scale adoption, they must be resource-efficient and seamlessly integrate into clinical workflows. Lightweight models developed through pruning, knowledge distillation, and edge AI can support real-time use in low-resource settings. At the same time, regulatory approvals, usability testing, and collaboration with clinicians are necessary to ensure safety, trust, and clinical acceptance.

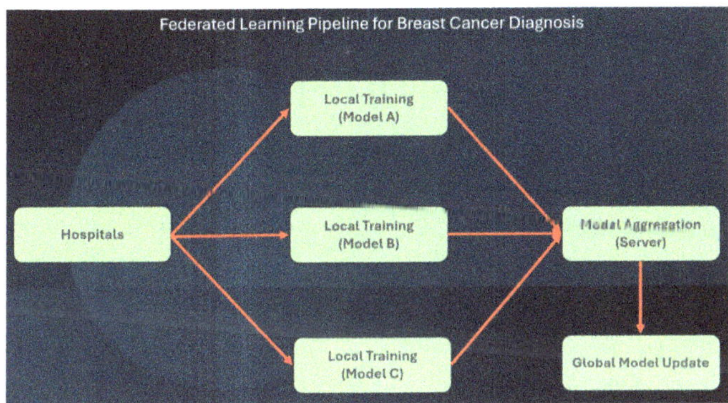

Figure 8.6: Federated learning pipeline for privacy-preserving breast-cancer diagnosis.

In summary, future directions in AI-based breast-cancer detection should prioritize inclusivity, transparency, efficiency, and privacy. By addressing these aspects, AI has the potential to become a reliable decision-support tool, complementing clinical expertise and improving patient outcomes.

8.8 Conclusion

Artificial intelligence has emerged as a transformative tool for breast-cancer detection and diagnosis. This chapter reviewed key deep-learning approaches, ranging from convolutional neural networks (CNNs) to Transformer-based architectures, and examined their applications across mammography, ultrasound, MRI, and histopathology imaging. Benchmark datasets were summarized, representative case studies were highlighted, and a comparative analysis of models was presented to provide a comprehensive overview of the field. Despite impressive progress, challenges such as limited data diversity, lack of interpretability, and barriers to clinical integration remain. Current solutions often excel in controlled research environments but face difficulties in real-world deployment, particularly across varied clinical settings and patient populations.

Looking ahead, AI-based breast cancer detection systems must evolve towards more inclusive, explainable, and resource-efficient frameworks. Emerging directions such as multi-modal learning, federated and privacy-preserving approaches, and lightweight deployment architectures offer promising pathways. Ultimately, the true potential of AI will be realized only through close collaboration with clinicians and healthcare stakeholders, ensuring these technologies complement medical expertise and contribute to earlier detection, better treatment planning, and improved patient outcomes.

9 Machine learning-based solutions for automated fracture detection from X-ray images

Abstract: This chapter examines how recent learning-based methods can support fast, consistent triage and reporting in routine practice. Accurate identification and grading of fractures on radiographs remain central to diagnosis and treatment planning; however, automation still contends with heterogeneous image quality, shifts between datasets, and the wide variety of fracture morphologies encountered across patient populations. Drawing on case studies and comparative evaluations, the chapter considers both algorithmic limits (e. g., performance drift outside the development domain, sensitivity to labeling noise, and class imbalance) and workflow frictions that arise during clinical integration. We synthesize design lessons with three recurring themes. First, disciplined data curation and preprocessing—standardizing projection views, exposure, and resolution—are prerequisites for stable performance. Second, close involvement of clinical experts in objective setting, error analysis, and acceptance criteria improves relevance and reduces overfitting to nonclinical artifacts. Third, explainable-AI tools (e. g., case-level rationales and region highlights) are needed so model decisions can be audited and trusted by readers. The chapter also reviews common evaluation practices, emphasizing sensitivity, specificity, and AUC for detection, as well as multicenter validation to quantify generalizability. Finally, we outline practical paths forward: robust training on diverse, multi-institutional cohorts; domain adaptation to mitigate site and protocol differences; strategies for scarce or skewed labels; and privacy-preserving collaboration. Taken together, these elements aim to pair stronger generalization with clearer interpretability, enabling automated systems to deliver consistent, dependable results across scanners, institutions, and clinical workflows.

Keywords: Automated fracture detection, X-ray image analysis, Machine learning in medical imaging, Deep learning for bone fractures, Clinical image variability, Explainable artificial intelligence (XAI)

9.1 Introduction

Bone fractures are a common medical condition resulting from traumatic injuries, repetitive stress, or underlying disorders such as osteoporosis. If left undetected or untreated, fractures may severely impair mobility and reduce quality of life. Accurate detection and classification–covering both the type and location of a fracture—are critical for appropriate treatment planning. Fractures may present in diverse forms, such as simple fractures with a clean break, comminuted fractures with multiple fragments, and subtle hairline fractures that are often difficult to detect (Table 9.1). Delayed or

https://doi.org/10.1515/9783111389059-009

Table 9.1: Common fracture types and associated diagnostic challenges.

Fracture Type	Description	Diagnostic Challenge
Simple	A clean break of the bone into two parts	Typically visible, but misinterpretation is possible in poor-quality images
Comminuted	Bone breaks into multiple fragments	Complex to classify due to irregular shapes and overlapping fragments
Hairline	Thin, subtle crack in the bone	Often missed because of low contrast or overlap with anatomical structures

incorrect diagnosis can lead to complications, including malunion, chronic pain, and long-term disability [125].

X-ray imaging continues to be the most widely used technique for diagnosing fractures because it is affordable, noninvasive, and capable of generating detailed skeletal images. The process works by transmitting electromagnetic radiation through the body, where dense tissues such as bones absorb more radiation, resulting in a clear depiction of skeletal structures on the radiograph. This enables identification of cortical disruptions, misalignments, and fracture patterns [31]. However, X-rays also have limitations. Small or low-contrast fractures, particularly hairline fractures, may remain undetected, especially when obscured by overlapping anatomical structures. Variability in image quality caused by equipment differences, patient positioning, and imaging protocols can further hinder interpretation [248]. These limitations highlight the need for advanced diagnostic methods that can improve the accuracy and reliability of fracture detection.

Automated detection of fractures in X-ray images is increasingly viewed as a practical way to support clinical decision-making, but building reliable systems remains a challenge. Considerable variation in imaging devices, patient anatomy, and acquisition settings often introduces inconsistencies that models must handle. The wide range of fracture presentations, from simple transverse breaks to more complex comminuted patterns, adds additional difficulty [7]. Early Computer-Aided Detection (CAD) approaches made use of handcrafted descriptors such as texture, edge profiles, and shape-based cues. Although these methods achieved some success, they typically required substantial manual input and struggled to generalize across datasets [345]. More recent work in deep learning, particularly with convolutional neural networks (CNNs), has shifted the field toward automated feature learning. Instead of relying on manually defined characteristics, CNNs extract hierarchical features directly from X-rays, improving accuracy and reducing the dependence on engineered features.

9.2 Automated bone fracture-detection from X-ray images

The task of automatically detecting bone fractures in X-ray images has become a growing focus of research, largely due to its promise to improve diagnostic accuracy, reduce errors, and make clinical workflows more efficient [295]. With recent progress in deep learning and medical image analysis, algorithms are now able to capture subtle and complex fracture patterns that radiologists might overlook, particularly in busy clinical settings. In the following discussion, attention is given to key preprocessing strategies, machine learning methods, and commonly used network architectures, together with examples from applied studies.

9.2.1 Image preprocessing techniques for enhancement

Preprocessing is an essential step in automated fracture detection because it conditions radiographs before they are analyzed by learning-based models. Common steps include contrast adjustment, noise reduction, and intensity normalization. Adaptive Histogram Equalization (AHE), particularly in its contrast-limited form (CLAHE), is often applied to highlight fine fracture lines and subtle discontinuities. Noise can be suppressed with filters such as Gaussian or median operators, provided that edge details important for identifying fractures are preserved. Standardizing images to a uniform size and intensity scale further improves training stability and evaluation consistency. Previous studies identify CLAHE as especially useful for improving the visibility of bone structures in X-ray segmentation and detection tasks [111].

9.2.2 Machine learning-based methods for fracture detection

Machine learning has enabled scalable and accurate analysis of radiographs for fracture detection. The following subsections summarize representative approaches.

9.2.2.1 Convolutional neural networks (CNNs)

CNNs are central to most fracture detection systems because they learn hierarchical spatial features directly from images. They can identify cortical breaks, discontinuities, and displacement patterns linked to fractures.

- *ResNet for wrist fractures*: Residual connections mitigate vanishing gradients, enabling deeper models to capture subtle fracture cues with high sensitivity and specificity [293].
- *Customized CNNs for long bones*: Architectures incorporating edge- or texture-focused layers improve recognition of spiral and comminuted fractures [132].
- *Lightweight CNNs*: MobileNet-like backbones make possible real-time inference in resource-constrained clinics, while maintaining diagnostic accuracy [331].

9.2.2.2 Transfer learning

Transfer learning addresses the scarcity of annotated datasets by fine-tuning models pre-trained on large image collections. This strategy reduces training time and improves generalization.

- *DenseNet on wrist radiographs*: Dense connectivity improves feature reuse and propagation, enhancing accuracy for subtle fracture cases [132].
- *EfficientNet for long bones*: Compound scaling provides a balance of performance and efficiency, suitable for high-throughput diagnostic settings.
- *ResNet on pediatric cohorts*: Pretrained ResNet models fine-tuned on pediatric data reported improved sensitivity and specificity by leveraging ImageNet features [293].

9.2.2.3 Hybrid models

Hybrid models integrate CNNs with sequence learners to leverage information from multiple projections or sequential X-rays.

- *CNN–RNN*: CNNs extract per-view features, while RNNs capture relationships across views, improving long-bone fracture detection [331].
- *CNN–BiLSTM*: Bidirectional recurrence enhances sensitivity for overlapping or subtle fractures [132].
- *CNN–GRU*: Provides faster inference than LSTMs while retaining sequential modeling capacity, useful in trauma settings.

9.2.2.4 Ensemble models

Ensemble learning combines outputs of multiple models to improve robustness and accuracy. This mitigates overfitting and increases generalizability.

- Weighted ensembles helped address class imbalance, improving recognition of rare fracture types [132].
- Stacking ensembles combining deep learning and traditional ML methods achieved superior accuracy on datasets with high variability [293].
- Recent staged ensembles applied YOLOv5 variants for fracture localization followed by EfficientNet-B3 classifiers, reporting AUC of 0.82 and accuracy of 0.81 on distal radius fractures [179].

9.2.2.5 Reinforcement learning approaches

Reinforcement learning, though less common, has been explored to optimize localization and classification tasks.

- Reinforcement learning optimized bounding box predictions for fracture localization, achieving precise delineation of complex cases [132].
- Policy-gradient approaches combined with CNNs prioritized high-probability regions dynamically, lowering misclassification rates.

9.2.3 Representative network architectures

Progress in network architectures has been crucial for automated fracture detection. The following families are widely applied:

- *Modern CNNs*: ResNet with skip connections supports very deep networks [293]; InceptionNet extracts both fine detail and broad patterns; DenseNet improves efficiency through dense feature reuse [132].
- *Region-based CNNs (R-CNN, Fast R-CNN, Faster R-CNN)*: These models focus on candidate regions before classification, achieving strong localization and accuracy for fractures [95].
- *YOLO*: A single-shot detector that enables real-time, grid-based localization and classification. Studies report sharp bounding boxes around fracture sites even with limited computational resources [95].
- *U-Net and variants*: Encoder–decoder structures with skip connections provide accurate pixel-level segmentation of fracture regions. Attention U-Net further emphasizes clinically relevant regions, enhancing performance in noisy datasets [132].
- *Ensemble architectures*: Multi-model ensembles (e. g., MobileNetV2, VGG16, InceptionV3, ResNet50) have demonstrated improvements in precision, recall, and F1 score for humerus and wrist fractures [285].

Overall, convolutional networks remain the dominant backbone for fracture detection. Transfer learning enhances performance when labeled data are limited, while hybrid and ensemble approaches improve robustness and adaptability. Region-based detectors and U-Net architectures are particularly valuable for localization and segmentation. Reinforcement learning remains exploratory but demonstrates potential for refining focus and optimizing predictions. Collectively, these approaches aim to deliver faster, more accurate, and scalable fracture-detection systems for clinical use. Table 9.2 presents the

Table 9.2: Representative approaches for automated fracture detection from X-ray images.

Method	Example Model	Application	Key Strength	Ref.
CNNs	ResNet	Wrist radiographs	High sensitivity to subtle cortical breaks	[293]
Region-based CNNs	Faster R-CNN	Pediatric fractures	Joint localization and classification	[95]
Transfer Learning	DenseNet, ResNet	Long bones, pediatrics	Improved generalization with limited data	[132, 293]
Hybrid Models	CNN–BiLSTM	Multiview radiographs	Captures inter-view dependencies	[331]
Ensemble Models	YOLO + EfficientNet	Distal radius fractures	Robustness, higher AUC	[179]
Reinforcement Learning	CNN + Policy Gradient	Fracture classification	Region-focused optimization	[132]

summary of a few relevant methods for automated fracture detection from X-ray images.

9.3 Performance analysis of bone fracture detection-methods

Rigorous evaluation is essential before automated fracture detection models can be relied upon in clinical workflows. While most studies report benchmark scores on shared datasets, performance metrics alone are insufficient. It is equally important to analyze the strengths and weaknesses of each approach—its sensitivity and specificity, robustness to noise and imaging variations, and scalability across sites and scanners.

Convolutional neural networks (CNNs) remain the dominant approach. By learning features directly from radiographs, CNNs have achieved sensitivity and specificity above 90 % in several studies on wrist- and long-bone fractures, highlighting their potential as clinical decision-support tools [125]. Nevertheless, performance often declines when models are transferred across institutions due to variations in patient populations, imaging equipment, and acquisition protocols.

Region-based detectors such as Faster R-CNN integrate feature extraction and region proposal stages in a unified framework. They have shown strong results in identifying complex fracture patterns, but their high computational cost can hinder real-time use [345]. To mitigate limited data availability, transfer learning with pretrained backbones such as ResNet and DenseNet has been widely adopted. This strategy improves efficiency, enhances sensitivity for subtle fractures, and increases cross-site adaptability [7]. Hybrid designs extend this further by combining CNNs with sequence models such as RNNs or LSTMs, enabling models to incorporate contextual cues from multiple views or sequential radiographs. While these approaches enhance performance, they often require substantial computational resources. Ensemble learning further boosts robustness by integrating predictions of multiple models. For example, hybrid ensemble frameworks that combine CNN-based deep learning models with traditional machine learning classifiers (e. g., Support Vector Machines) have reported significant gains in accuracy and sensitivity for challenging fracture cases [33]. However, such frameworks add complexity and increase inference cost.

Table 9.3 summarizes reported performance trends for representative methods in fracture detection.

9.3.1 Performance evaluation metrics

Assessing clinical readiness requires standardized, quantitative measures. Beyond accuracy, metrics such as sensitivity, specificity, and AUC reveal how well models detect fractures and avoid false alarms. Together, these metrics provide a more complete assessment than accuracy alone.

Table 9.3: Performance trends of representative fracture-detection methods. Reported values are from individual studies using multiple datasets and protocols; results of direct comparisons should therefore be interpreted with caution.

Method	Dataset/Setting	Sensitivity	Specificity / AUC	Ref.
CNN (ResNet)	Wrist radiographs (clinical)	>90 %	>90 %	[125]
Faster R-CNN	Pediatric radiographs	High	Good localization; high compute cost	[345]
Transfer Learning (DenseNet, ResNet)	Long bones, pediatric sets	>85 %	AUC >0.95	[7]
Hybrid CNN–RNN	Multiview series	Improved	Robust to complex cases	[132]
Ensemble (CNN + SVM)	Complex fracture datasets	Increased	Higher sensitivity, stable accuracy	[33]

9.3.1.1 Key metrics

- *Accuracy:* Proportion of correctly classified cases (fractured and nonfractured).

$$\text{Accuracy} = \frac{\text{TP} + \text{TN}}{\text{TP} + \text{TN} + \text{FP} + \text{FN}}$$

- *Sensitivity (Recall):* Proportion of true fracture cases correctly detected.

$$\text{Sensitivity} = \frac{\text{TP}}{\text{TP} + \text{FN}}$$

- *Specificity:* Proportion of nonfracture cases correctly identified.

$$\text{Specificity} = \frac{\text{TN}}{\text{TN} + \text{FP}}$$

- *F1 Score:* Harmonic mean of precision and recall, balancing false negatives and positives.

$$\text{F1 Score} = 2 \times \frac{\text{Precision} \times \text{Recall}}{\text{Precision} + \text{Recall}}$$

- *AUC-ROC:* Threshold-independent summary of a model's ability to discriminate between fracture and nonfracture cases.
- *Mean Average Precision (mAP):* Widely used in detection tasks; averages precision across recall levels.

Recent studies report accuracy above 90 % and sensitivities exceeding 85 % for CNN-based systems on wrist and long-bone datasets [125]. Transfer learning approaches further improve generalization, with AUC values above 0.95 in several cases [132]. Ensemble strategies provide additional gains in F1 score by addressing class imbalance, reducing both false positives and negatives [33].

9.3.2 Presentation of experimental results

Experimental validation is critical to understanding how models behave across different datasets, imaging conditions, and protocols. Preprocessing steps such as contrast enhancement and denoising improve faint fracture visibility, boosting performance on external data [223]. Representative findings include:

- *Data augmentation for robustness:* Random shifts, rotations, and rescaling improve generalization across imaging conditions [90].
- *Transfer learning plus preprocessing:* Pretrained backbones fine-tuned with preprocessing yield high sensitivity and precision in skeletal bone-age and fracture detection tasks [269].
- *Scalability across tasks:* Comparative studies on multi-label chest X-rays demonstrate architectures that generalize well across targets, informing fracture detection research [30].

9.3.3 Effectiveness and limitations

Deep learning methods have brought notable improvements to automated fracture detection, with studies reporting high sensitivity, strong specificity, and even the possibility of real-time use in certain applications. Despite these advances, a number of important limitations remain:

- *Class imbalance:* In many datasets, nonfracture cases greatly outnumber positive examples. This imbalance can bias training and reduce the model's ability to detect fractures reliably [125].
- *Dataset variability:* Imaging differences across institutions—such as scanner type, acquisition protocol, or resolution—often reduce generalizability and lead to performance drops on external data [114].
- *Interpretability:* Many deep models operate as opaque systems, making it difficult for clinicians to understand why a particular decision was made. This lack of transparency continues to limit trust and slow adoption [132].
- *Computational complexity:* Training and inference can be resource intensive, which restricts the use of advanced models in real-time workflows or in settings with limited hardware [90].
- *Data privacy and security:* Because medical images are highly sensitive, collaborative or multi-site training must also address regulatory and ethical constraints (e. g., HIPAA, GDPR), which complicates access to large and diverse datasets.

Future directions: Addressing these limitations requires:
- Domain adaptation and federated learning to enable robust cross-site generalization while preserving data privacy.
- Explainable AI (XAI) techniques, such as saliency maps and attention heatmaps, to improve transparency and clinical trust.
- Lightweight models (e. g., knowledge distillation, pruning, quantization) to support real-time use on standard hardware.
- Advanced augmentation strategies, including GAN-based synthetic images and adaptive resampling, to mitigate class imbalance.
- Standardized imaging protocols and shared benchmark datasets for reproducibility and fair comparison across studies.
- Consideration of regulatory validation and approval as part of clinical readiness assessment.

9.4 Challenges in automated bone fracture-detection: insights from case studies

Research on building automated systems for detecting bone fractures in X-ray images has revealed a number of recurring difficulties, many of which are well documented in case studies. These reports describe a mix of technical, clinical, and operational issues that arise when artificial intelligence (AI) is applied to medical imaging. Common problems include the quality of available data, the ability of models to generalize across institutions, and the challenge of explaining predictions to clinicians. In addition, combining fracture detection with other clinical parameters for a more holistic analysis remains a demanding task.

One significant study explored the application of deep supervised learning for bone-fracture detection, emphasizing the challenge of data annotation [173]. Annotated datasets of high quality are essential for training deep-learning models; however, annotation is a labor-intensive process requiring domain expertise. The study underscored how imbalanced datasets, in which fractured cases are significantly outnumbered by nonfractured ones, skew model predictions and lead to poor performance on minority classes. Subtle fractures, in particular, were difficult to annotate and were often missed during training.

Another study focused on segmentation challenges in X-ray images, a crucial preprocessing step for fracture detection [263]. While this study primarily addressed segmentation in dental X-rays, the challenges it identified are equally relevant to bone-fracture detection. Overlapping anatomical structures, noise, and low contrast in X-ray images often hinder the accurate segmentation of bone regions, negatively affecting model performance. Poor segmentation propagates errors through the detection pipeline, emphasizing the need for robust preprocessing algorithms.

Generalizability was evaluated in a hip-fracture study [81]. The model matched radiologist performance on the internal (training) set, but accuracy declined on external datasets. Likely reasons included changes in imaging protocols, differences in patient mix, and variable X-ray quality across sites. This highlights the necessity of training and validating models on diverse, multi-site datasets to ensure robustness outside controlled environments.

Hsieh et al. [104] paired fracture detection with bone mineral density (BMD) prediction. This demonstrated how a single system can detect injury, while also providing clinically useful context. Although results were promising, the approach required large datasets and substantial computational resources, which are difficult to secure in resource-limited settings. Finally, interpretability remains essential for deployment: Clinicians require clear, case-level explanations of predictions before relying on automated models in routine care.

9.5 Conclusion and key insights

Synthesizing the evidence from case studies and performance analyses, several important lessons emerge that can guide the development of robust and clinically relevant automated bone fracture-detection systems. Case studies consistently highlight three fundamental practices. First, ensure data quality: Clear labeling and balanced representation of fractured and nonfractured films are essential to avoid bias toward the majority class. Second, adopt efficient and consistent labeling strategies—using streamlined tools, standardized annotation protocols, and, when possible, dual review—to expand datasets without compromising quality. Third, design training pipelines that reflect real-world diversity: Incorporating multiple scanners, exposure settings, and views (AP, lateral, oblique) improves generalizability and prepares systems for the variability encountered in clinical environments. Together, these practices address the recurring issues of class imbalance and protocol variability, thereby supporting more stable model performance [173].

Segmentation challenges, as noted in prior studies, underline the importance of advanced algorithms capable of handling noise, overlapping structures, and anatomical variability [263]. Robust preprocessing techniques that improve segmentation accuracy are essential for enhancing the overall detection pipeline. The issue of generalizability, highlighted by Gale et al., emphasizes the need for training on datasets representative of diverse imaging protocols and patient populations [81]. Methods such as transfer learning and domain adaptation can support cross-institutional performance and reduce overfitting to single-site datasets. Finally, the work of Hsieh et al. [104] demonstrates the potential of integrating multiple clinical parameters, such as bone mineral density (BMD), to create multipurpose AI systems. However, ensuring interpretability remains critical. Developing transparent models that provide clear reasoning for predictions will strengthen clinician trust and support routine adoption.

In conclusion, the lessons drawn from this work point to several recurring challenges that need to be resolved before such systems can be safely used in clinical practice. Among the main priorities are the creation of reliable annotated datasets, the design of preprocessing approaches that can cope with imaging variability, strategies to achieve generalizability across institutions, and the development of models whose predictions can be clearly explained to clinicians. Progress in these areas will strengthen diagnostic accuracy and, equally important, help advance AI-based fracture detection toward routine clinical use.

10 Methods for Alzheimer's disease detection using computer-aided diagnosis (CAD) systems

Abstract: Alzheimer's disease (AD) is a growing public-health concern, with cases expected to rise steeply in the coming decades. This chapter reviews recent advances in computer-aided diagnosis (CAD) aimed at earlier detection of AD and clinical decision support. Conventional approaches—clinical examinations, structural magnetic resonance imaging, PET, EEG, and cerebrospinal fluid biomarkers—offer important signals, but often miss prodromal changes. Newer pipelines combine convolutional and 3D convolutional neural networks, recurrent modules such as FSBi-LSTM, and multi-source feature fusion spanning MRI, PET, CSF, and genetics. Unsupervised clustering, texture descriptors, and gray-level co-occurrence features complement these models by capturing subtle structural and functional patterns. We also examine explainable AI techniques (e. g., Grad-CAM, LIME, SHAP), stronger preprocessing workflows, and generative augmentation strategies that mitigate class imbalance. In the future, priorities include robust multimodal fusion, lightweight architectures based on transfer learning, and cost-effective hardware accelerators for point-of-care use. Close collaboration between clinicians and AI researchers is essential to ensure reliability, interpretability, and ethical implementation. Together, these developments indicate a maturing CAD ecosystem with the potential to support a faster and more accurate diagnosis of AD, ultimately improving patient outcomes.

Keywords: Computer-aided diagnosis, Alzheimer's disease detection, Medical image analysis, MRI

10.1 Introduction

Alzheimer's disease (AD) is one of the most serious public health concerns worldwide. Characterized by a gradual decline in memory and cognitive function, it has a profound impact not only on patients, but also on families, creating emotional and financial burdens. According to the World Health Organization (WHO), nearly 50 million people currently live with AD and this number is projected to triple by 2050 [299]. With the global population aging, the prevalence of AD is expected to increase dramatically, underscoring the urgent need for effective diagnostic approaches and intervention strategies. This chapter reviews conventional and emerging diagnostic methods, with particular attention to the role of artificial intelligence (AI) in enabling early-stage detection.

AD is a neurodegenerative disorder that causes progressive neuronal loss. The disease begins slowly and worsens over time, with early symptoms often including short-term memory loss. As the condition progresses, patients may develop mood disturbances, language difficulties, impaired self-care, and other neurological deficits that

https://doi.org/10.1515/9783111389059-010

eventually affect vital body functions, leading to death. Although therapy currently does not stop or reverse the progression of Alzheimer's disease, identifying it early can still make a substantial difference. Detection in time allows therapeutic interventions that can slow down and, in turn, improve the patient's quality of life. However, achieving this remains difficult. Conventional diagnostic practices often fail to capture mild cognitive impairment (MCI), the stage where intervention is most beneficial. Hereditary factors also play a decisive role, with almost 80 % of cases showing genetic links; families with a history of AD are at increased risk, underscoring the need for early and proactive detection [42, 271].

Traditionally, clinicians have relied on a combination of neurological assessments, neuropsychological tests, and imaging studies. Techniques such as structural magnetic resonance imaging (sMRI), positron emission tomography (PET), and functional MRI (fMRI) have been critical for identifying structural and functional brain changes [299]. In parallel, biomarker analyses of cerebrospinal fluid (CSF) and blood samples provide complementary information on amyloid beta and tau protein concentrations [297]. Cognitive screening tools, including the Mini Mental State Examination (MMSE), remain widely used to evaluate memory and executive function, although they also face limitations in detecting subtle early-stage impairment [206]. However, each method has limitations, particularly in the detection of early disease, where diagnostic accuracy is often insufficient. Additional challenges include procedural risks, high costs, and inconsistent associations between genetic markers and early-onset AD.

Traditional diagnostic approaches face several limitations:
- *Cognitive assessments:* While inexpensive and easy to administer, they are prone to subjectivity and lack sensitivity for early-stage detection.
- *Neuroimaging:* Standard imaging often fails to capture early pathological changes during the MCI phase.
- *PET scans:* Although highly informative, they are costly and require specialized infrastructure, limiting their accessibility.
- *Multimodal data:* Interpreting multimodal input is computationally demanding and often delays diagnosis.

Furthermore, clinical symptoms often emerge only after pathological changes have already begun. EEG studies have shown abnormal slow-wave rhythms in AD patients, suggesting EEG as a potential noninvasive biomarker [120]. Nevertheless, immature diagnostic frameworks and insufficient data integration can still lead to misclassification. A comparative overview of diagnostic methods is provided in Table 10.1.

Table 10.1: Comparison of diagnostic methods for Alzheimer's disease.

Method	Advantages	Disadvantages
Cognitive Assessments	Low-cost; simple to administer	Subjective; low sensitivity in early stages
sMRI	Noninvasive; structural brain insights	Expensive; limited early-stage sensitivity
PET	Enables biomarker visualization	High cost; limited availability
AI-based Models (e. g., CNNs)	High sensitivity; supports multimodal integration	Requires large datasets; interpretability challenges

10.1.1 The role of AI and emerging technologies

Machine learning (ML) and AI have reshaped approaches to AD detection, particularly in the analysis of neuroimaging data. Deep learning methods such as Convolutional Neural Networks (CNNs) have demonstrated strong performance in detecting subtle structural changes in brain regions, including the hippocampus and entorhinal cortex [235]. Recent advances also highlight lightweight networks for real-time applications, combining diagnostic accuracy with computational efficiency. Despite these developments, hardware-driven innovations remain underexplored and could offer new opportunities for clinical translation.

Other techniques include weighted hypergraph convolutional networks (WHGCN), AdaBoost ensemble methods, and High-Order Brain Functional Networks (HO-BFN), each contributing unique perspectives [349]. MRI-based models such as Conv-Swinformer and MPS-FFA have added further insight, though challenges such as data imbalance, high computational cost, and limited interpretability persist. Their complementarity is illustrated in Fig. 10.1.

Early diagnosis of AD has important implications for patients, caregivers, and healthcare systems. Detecting the disease during the MCI stage can:
- enable treatments aimed at slowing cognitive decline,
- allow participation in clinical trials for disease-modifying therapies, and
- support the design of personalized care plans that improve quality of life.

As shown in Fig. 10.2, early detection also benefits public health by enabling better resource planning and policy development. Advances in signal processing, such as converting EEG signals from the time domain to the frequency domain using the Fast Fourier Transform (FFT), further improve diagnostic precision when integrated with neural networks.

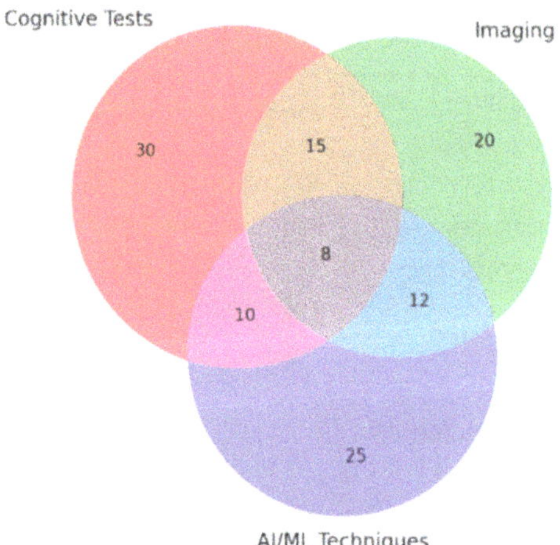

Figure 10.1: Venn diagram illustrating complementarity among diagnostic approaches.

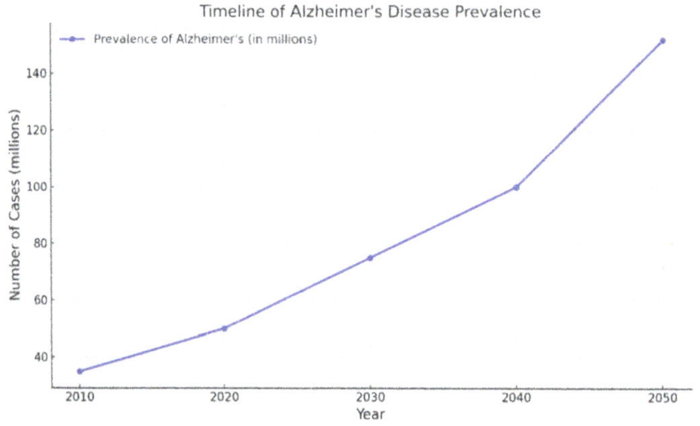

Figure 10.2: Projected global increase in AD prevalence, underscoring the urgency of early diagnosis and intervention strategies.

10.1.2 OASIS dataset description

The OASIS (Open Access Series of Imaging Studies) dataset is a widely used benchmark for Alzheimer's research [119]. It contains imaging data from 416 subjects aged 18 to 96 years, including T1 and T2-weighted magnetic resonance modalities. Classification labels include Normal Control (NC) and Alzheimer's Disease (AD). Key components include:

- *Cross-sectional data:* 100 participants aged 60+ with mild to moderate AD and 416 control participants.
- *Longitudinal data:* 373 sessions from 150 participants, including cases transitioning from nondemented to demented status.

10.1.3 Challenges

Despite notable advances, important gaps remain in the detection of AD.

- *Data heterogeneity:* Variability in imaging protocols and clinical data reduces the generalizability of the model
- *Interpretability:* Many deep learning models function as 'black boxes', limiting transparency and clinical acceptance.
- *Ethical considerations:* Issues related to privacy, algorithmic bias, and equitable access to AI-based diagnostics remain unresolved.

The preprocessing of neuroimaging data also requires complex pipelines, while diagnostic criteria differ between institutions, complicating model validation [135]. sMRI continues to demonstrate potential as a noninvasive biomarker, with hippocampal volume, cortical thickness, and overall brain atrophy strongly linked to disease progression. AI methods such as CNNs and Vision Transformers (ViTs) have proven effective in extracting and analyzing these features, while interpretability tools such as Grad-CAM enhance usability in clinical practice. Alternative approaches involving texture analysis, complex networks, and motor imagery (MI) signals also provide promising directions for early detection [66].

10.2 Related work

Research on Alzheimer's disease (AD) diagnosis has expanded significantly in the past decade, with contributions that span conventional clinical studies, neuroimaging-based methods, and advanced AI-driven frameworks. This section reviews representative approaches, focusing on convolutional neural networks (CNNs), 3D-CNN and recurrent hybrid frameworks, multimodal feature fusion, IoT-based solutions, and unsupervised approaches. The key strengths, limitations, and remaining challenges are highlighted.

10.2.1 CNN-based computer-aided diagnosis (CAD)

Convolutional neural networks (CNN) have become central to computer-aided diagnosis (CAD) systems for AD. When applied to ^{18}FDG-PET images from the ADNI dataset, CNN-based approaches have achieved accuracies and sensitivities that exceed 94 % [91]. Automated feature extraction enables efficiency and strong diagnostic performance, although challenges remain in terms of limited dataset availability, lack of longitudinal evaluation, and insufficient validation in diverse populations. Improving generalizability and systematically comparing CNNs with alternative methods are ongoing priorities.

Transfer learning (TL) has further enhanced CNN performance, particularly when applied to sagittal magnetic resonance data. TL makes possible robust classification even with limited datasets, making it highly suitable for domains where labeled medical data is scarce. Studies report that TL-sagittal MRI achieves performance comparable to horizontal-plane MRI [218]. However, gaps remain in clinical validation and direct comparison across acquisition planes, limiting clinical adoption.

The general architecture of CNN (Fig. 10.3) mirrors the hierarchical structure of the human visual cortex.

- *Convolutional layers:* extract features using kernel filters,
- *Pooling layers:* reduce dimensionality, while preserving key information, and
- *Fully connected layers:* map features to classification outputs.

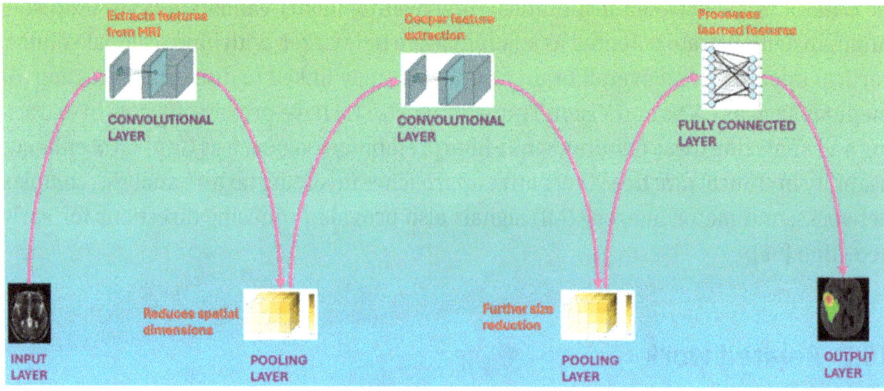

Figure 10.3: Typical CNN architecture used for AD detection, showing convolution, pooling, and classification stages.

10.2.2 3D-CNN and FSBi-LSTM-based frameworks

Building on 2D CNNs, 3D CNNs (3D-CNNs) capture spatial and volumetric information from MRI and PET data. When combined with FSBi-LSTM, these models can incorporate temporal patterns of disease progression. Feng et al. [76] demonstrated that such a framework effectively distinguished between progressive MCI (pMCI), stable MCI (sMCI), AD, and normal controls (NC) using the ADNI dataset, outperforming several existing models.

Despite their strong performance, these models face the challenges of interpretability and computational burden. Limited insight into the characteristics driving classification hinders clinical trust, while the exclusion of longitudinal temporal dynamics restricts full characterization of disease progression. In addition, issues of data privacy, potential bias, and ethical considerations remain key obstacles to a wider deployment.

10.2.3 Feature-fusion strategies

Multimodal frameworks integrate MRI, PET, cerebrospinal fluid (CSF), and genetic biomarkers to capture complementary pathological information. Feature-fusion strategies, such as simple concatenation, graph-based integration, or attention-driven mechanisms, have significantly improved classification accuracy [275]. Recent transformer-based approaches (e. g., ViT, VL-BERT) provide global contextual modeling, enabling deeper cross-modal interactions [70].

Although these methods demonstrate strong potential, they often require extensive pretraining and substantial computational resources, which can limit their practical applicability in clinical environments. Ongoing efforts aim to optimize transformers for medical imaging, balancing accuracy with efficiency.

Deep learning has also facilitated advances in clustering and representation learning. DenseNet-based frameworks and sparse autoencoders improve MRI feature learning and have shown competitive performance in distinguishing AD from normal controls [6, 43]. However, overfitting, dataset specificity, and interpretability remain persistent challenges.

10.2.4 IoT-based and real-time frameworks

Beyond neuroimaging, IoT-driven frameworks leveraging fog-cloud architectures have been proposed for real-time monitoring of cognitive decline. These systems collect multimodal data from both invasive and noninvasive sensors, analyzed through CNN, RNN, and LSTM models [29]. Such approaches offer opportunities for continuous patient monitoring and personalized intervention. However, practical deployment faces obstacles such as high energy consumption, latency, and sensitivity to hyperparameter tuning. Research is ongoing to optimize IoT architectures for scalability and heterogeneous data integration.

10.2.5 Handling class imbalance

One persistent challenge in Alzheimer's disease (AD) datasets is class imbalance, where AD cases often represent only a small fraction of the data. To address this, researchers frequently apply the synthetic minority oversampling technique (SMOTE), which generates additional samples of the underrepresented class and helps improve classification reliability [61]. However, while this strategy can enhance balance, it must be applied cautiously: Excessive oversampling can introduce artificial patterns, leading to overfitting and biased model evaluations.

10.2.6 Unsupervised learning for Alzheimer's disease detection

Although supervised learning frameworks—such as CNNs, hybrid models, and multimodal fusion—have achieved notable progress, their reliance on large annotated datasets and computationally heavy pipelines limits broader applicability. As a complementary direction, researchers have explored unsupervised approaches that emphasize clustering and segmentation tasks without the need for extensive manual labeling. These methods are particularly valuable in the preprocessing stage, where they help suppress noise and isolate relevant brain regions for later analysis.

Among these, segmentation-based methods such as k-means clustering have gained attention for their speed and ease of use when applied to large datasets. However, their simplicity can also be a drawback, often leading to incomplete identification of affected areas of the brain. To mitigate such shortcomings, hybrid pipelines have been proposed that combine unsupervised mapping with classification strategies. This integration enables efficient grouping of regions and more accurate categorization of Alzheimer's-related changes. Such systems typically include five stages: preprocessing, clustering, feature extraction, classification, and validation. The goals are to reduce iteration counts, compress feature vectors, minimize computational cost, and improve efficiency [282].

Image preprocessing further enhances the quality of MRI input by removing noise and irrelevant components such as the skull. For instance, median filters effectively suppress Gaussian and Poisson noise, while preserving edges critical for diagnosis. A comparison of commonly used filters is presented in Table 10.2.

Table 10.2: Comparison of denoising filters.

Filter Type	Noise Type Removed	Edge Preservation	Processing Time (ms)
Median Filter	Gaussian, Poisson	Excellent	Low
Bilateral Filter	Gaussian	Good	Moderate
Gaussian Filter	Gaussian	Poor	Low
Nonlinear Filters	Various	Moderate	High

Additional techniques, such as the Brain Surface Extractor (BSE), further reduce computational overhead by isolating essential brain regions, while removing nonrelevant structures like the skull, eyes, and scalp. This step streamlines the subsequent morphological analysis and improves the precision of AD-related feature extraction [139].

In general, the literature demonstrates that AI-driven methods have substantially advanced Alzheimer's disease detection. CNNs and their variants dominate supervised learning, offering high sensitivity and specificity, while 3D-CNN and hybrid RNN frameworks capture temporal and volumetric disease patterns. Multimodal fusion strategies extend diagnostic power by integrating MRI, PET, CSF, and genetic biomarkers, though computational costs remain a barrier. IoT-based frameworks broaden the scope to real-time monitoring, whereas techniques such as SMOTE address dataset imbalance. Finally, unsupervised methods, including clustering and advanced preprocessing pipelines, provide valuable support for segmentation and feature extraction when labeled data are scarce. Together, these approaches highlight both the promise of AI in clinical diagnosis of AD and the urgent need for solutions that are generalizable, interpretable, and efficient in real-world healthcare settings.

10.3 Feature extraction

Feature extraction plays a central role in the detection of Alzheimer's disease by isolating discriminative properties from MRI images that can be used for classification. One widely used method is the Gray-Level Co-Occurrence Matrix (GLCM), which captures both first-order and second-order texture features. These features describe not only the intensity of individual pixels but also the spatial relationships between pairs of pixels, offering robustness under scaling and rotation [170, 260].

To streamline analysis, regions of interest (ROIs) are often downsampled into 4×4 voxels from original 256×256 grayscale MRI images. This reduction in dimensionality accelerates computation while preserving critical information. The extracted properties include variance, homogeneity, entropy, angular moment, and correlation, each providing complementary information on structural brain changes associated with AD. The relevance of these characteristics is summarized in Table 10.3.

Table 10.3: Key GLCM-derived features and their relevance for AD detection.

Feature	Type	Relevance to AD Detection
Variance	Second-order	Highlights variations in pixel intensity.
Homogeneity	First-order	Identifies uniformity in grayscale regions.
Entropy	Second-order	Detects randomness and irregularities.
Correlation	Second-order	Measures pixel relationships over distance.
Angular Moment	First-order	Emphasizes pixel intensity concentration.

Once extracted, these texture features are typically used as inputs to classification models such as Support Vector Machines (SVMs), Random Forests, or deep neural networks. By transforming raw magnetic resonance data into compact and discriminative characteristic vectors, the models can distinguish more effectively between normal controls, patients with mild cognitive impairment, and those with Alzheimer's disease. This step forms a crucial bridge between image preprocessing and the subsequent machine learning or deep learning stages in the diagnostic workflow.

10.4 Future directions

Alzheimer's disease (AD) remains one of the most pressing challenges in modern healthcare, affecting millions of people around the world. Despite significant progress in diagnostic research, there are major limitations, including data scarcity, interpretability, high computational demands, and limited clinical integration. Future work should focus on strategies that combine technological innovation with medical expertise. This section outlines promising directions for advancing AD detection.

10.4.1 Enhanced use of multimodal imaging and data fusion

Integrating different imaging modalities, such as MRI, PET, CT, and fMRI, offers a way to strengthen the reliability of the Alzheimer's diagnosis. Since each technique provides unique information about the brain, from structural anatomy to metabolic activity and functional connectivity, while their combined use paints a fuller picture of the progression of the disease. Looking ahead, research should aim to refine fusion methods that can merge imaging data with genetic markers and clinical records in a single diagnostic framework. A system built on such multimodal integration could not only support earlier identification of individuals at risk, but also guide the creation of treatment plans tailored to the patient's specific condition [105].

10.4.2 AI and deep learning for robust diagnostics

Artificial intelligence (AI), particularly deep learning (DL), has demonstrated strong potential to automate AD diagnosis by capturing complex patterns in neuroimaging data. Unlike conventional methods that rely heavily on manual interpretation, DL models—especially convolutional neural networks (CNNs)—deliver reproducible and accurate results. Future research should emphasize refining the model sensitivity to subtle disease markers. Transfer learning (TL), where pretrained networks are adapted to AD data, offer a promising approach to mitigate data scarcity and enhance generalization. Furthermore, generative adversarial networks (GANs) can augment datasets by synthesizing realistic medical images, improving robustness, and reducing reliance on large annotated cohorts [87].

10.4.3 Explainable AI (XAI) for clinical adoption

A major barrier to the adoption of AI-based systems in clinical practice is their limited transparency. Clinicians require interpretable outputs to trust diagnostic recommendations. Future work should focus on Explainable AI (XAI) approaches such as Local Interpretable Model-Agnostic Explanations (LIME), SHapley Additive Explanations (SHAP), and Gradient-weighted Class Activation Mapping (Grad-CAM). These tools can provide case-level explanations by visualizing the characteristics that influence predictions, thus improving clinical acceptance and integration [143, 163, 246].

10.4.4 Improved data preprocessing and augmentation techniques

Reliable preprocessing pipelines that cover normalization, registration, segmentation, and skull removal are essential for consistent performance across data sets. Future directions include refining these steps to improve cross-dataset applicability and exploring advanced augmentation techniques. GAN-based augmentation is especially promising since it can generate high-quality synthetic images to mitigate data imbalance and expand training cohorts, thus improving the accuracy and robustness of the model [136].

10.4.5 Hardware implementation for real-time diagnosis

Beyond software models, hardware optimization is critical for practical deployment. Magnetic resonance imaging and PET provide detailed information, but remain resource intensive and costly. EEG and other low-cost modalities could complement or substitute large-scale imaging in certain settings. Research into specialized hardware accelerators and analog circuits could enable efficient real-time processing at the point of care. Such innovations would facilitate early diagnosis in routine clinical environments and expand access to AD detection technologies.

10.4.6 Clinical integration and collaboration with medical experts

For AI-based diagnostic tools to be clinically viable, they must align with healthcare workflows and meet the requirements of medical professionals. Future research should emphasize the close collaboration between AI developers and clinicians to ensure usability, interpretability, and regulatory compliance. Effective partnerships will help produce diagnostic systems that are both technically advanced and clinically practical, supporting widespread adoption in healthcare settings.

10.4.7 Extending to other neurological disorders

Although this chapter focuses on AD, similar approaches hold promise for other neurodegenerative diseases, including Parkinson's disease, multiple sclerosis, and Huntington's disease. Developing generalized frameworks that leverage multimodal data and advanced machine learning could enable cross-disease analysis, offering insights into both shared and unique patterns of neurological decline. Such an expansion would contribute to versatile diagnostic tools and could inform improved treatment strategies across disorders.

10.5 Conclusion

The landscape of Alzheimer's diagnosis is changing rapidly with the adoption of advanced imaging methods and artificial intelligence. Unlike traditional assessments that often miss early changes, the integration of structural and functional imaging with AI has made it possible to detect subtle patterns of neurodegeneration much earlier. These advances offer the potential for a timely diagnosis, closer patient monitoring, and ultimately better quality of care.

However, several hurdles remain. Models must be interpretable enough for clinicians to trust, flexible enough to generalize across diverse populations, and affordable enough for use outside of specialized centers. Addressing these issues will require cooperation between fields: Clinicians, AI scientists, and engineers must work together to design systems that are scientifically rigorous and clinically practical.

Key Lessons:

- Combining imaging with artificial intelligence improves early diagnosis, but practical deployment requires stable preprocessing and standardization.
- Multimodal integration of imaging, genetics, and clinical data can raise diagnostic accuracy, though it adds computational and logistical complexity.
- Deep learning and IoT-enabled monitoring show promise for real-time support, but face constraints such as latency, energy use, and hardware demands.
- Trust remains central; explainable AI techniques (e. g., Grad-CAM, SHAP) are essential for wider acceptance in clinical workflows.
- The long-lasting progress depends on interdisciplinary collaboration to ensure that new tools are not only powerful in theory but also feasible, affordable, and accessible in practice.

11 Transfer learning for medical image analysis in computerized disease diagnosis

Abstract: Transfer learning has become an increasingly valuable approach in medical image analysis, particularly in contexts where annotated data are limited and training models from scratch would be computationally prohibitive. This chapter examines the adaptation of pretrained convolutional neural networks and self-supervised learning techniques for a wide range of medical imaging applications, such as disease diagnosis, organ segmentation, and the detection of lesions. Emphasis is placed on practical implementations of transfer learning using established architectures such as VGG16, ResNet, DenseNet, and Inception, as well as on more recent developments in multistage and self-supervised pretraining strategies. In addition, the chapter explores the role of Vision Transformers (ViTs) and hybrid CNN-transformer models, which are emerging as powerful alternatives for extracting contextual and structural information from complex medical images. The chapter reviews the performance of these models across various imaging modalities (e. g., X-ray, CT, MRI, and ultrasound) and evaluates their effectiveness in real-world clinical scenarios. Special attention is paid to multimodal learning, where transfer learning enables the integration of complementary data sources (e. g., imaging, clinical records, and genomic data) to improve diagnostic performance. In addition, case studies are presented to illustrate how transfer learning accelerates deployment in resource-limited healthcare settings by reducing both training time and computational costs. Key challenges, such as domain shifts between natural and medical images, limited generalizability between patient populations, model interpretability, and data privacy constraints, are also discussed in detail. The difficulty of aligning pretrained features with domain-specific representations, particularly when transferring from nonmedical datasets like ImageNet, remains a critical bottleneck. Privacy-preserving approaches, including federated learning and differential privacy, are highlighted as promising strategies to mitigate concerns related to sensitive medical data sharing. By synthesizing current research and real-world case studies, this work highlights both the opportunities and limitations of transfer learning in advancing computer-aided diagnosis. Finally, future directions are outlined, including the development of domain-specific pretraining datasets, the incorporation of explainable AI methods to improve transparency, and the design of lightweight architectures for efficient clinical deployment. Together, these advances aim to make transfer learning-based models more robust, interpretable, and broadly applicable in clinical practice.

Keywords: Transfer learning, Medical image analysis, Deep learning, Computer-aided diagnosis, Self-supervised learning, Domain adaptation

https://doi.org/10.1515/9783111389059-011

11.1 Introduction

Recent progress in machine learning, especially deep learning, has revolutionized medical image analysis. Within this domain, transfer learning has proven to be a highly effective strategy to address the specific challenges involved in interpreting medical images such as X-rays, MRIs, CT scans, and ultrasound for computer-assisted disease diagnosis. By leveraging knowledge gained from large-scale datasets, transfer learning enables the adaptation of pretrained models to specialized medical imaging tasks, reducing both the need for extensive annotated datasets and the computational burden of training models from scratch.

This section sets the stage by explaining what transfer learning is, why it matters to machine learning practice, and how it has been used in medical imaging. We outline the model families that commonly underpin current systems: VGG, ResNet, DenseNet, and Inception, and we note newer entrants such as EfficientNet and Vision Transformers, which scale well and capture longer-range contexts. Brief examples show where transfer learning is routinely useful: lesion detection, tumor classification, organ segmentation, and even fracture assessment across multiple modalities. We also discuss why TL helps in practice (limited labels, high annotation cost, variability in acquisition protocols) and where it can struggle (domain shifts between natural and medical images, limited interpretability, privacy, and regulatory constraints). In general, the section argues that TL has been a catalyst for progress in medical image analysis, while demonstrating that further work on robustness, transparency, and clinically viable deployment is still needed.

11.1.1 Definition and significance in machine learning

Transfer learning refers to the reuse of knowledge acquired in one problem to accelerate or improve learning in a related problem [205]. In deep vision models, this typically means starting from convolutional neural networks (CNNs) that were pretrained on large image corpora (e. g., *ImageNet*) and adapting them to a clinical objective. A network that originally learned to recognize everyday objects can be fine-tuned to flag abnormalities in chest X-rays or CT scans. The intuition is well known: Early layers encode broadly useful patterns such as edges, textures, and simple shapes, while later layers are adjusted to the semantics of the target task.

The appeal in medical imaging is largely practical. High-quality annotations demand expert time (e. g., delineating tumors in mammograms or nodules in CT) that makes large labeled sets expensive and slow to assemble. On top of that, data-governance frameworks such as HIPAA and GDPR constrain wide sharing of clinical data. By initializing from strong, generic representations and then adapting a smaller portion of the model, transfer learning reduces the number of labels required, lowers the risk of overfitting on small cohorts, and shortens the path to a usable model. Transfer learning helps overcome these limitations by enabling effective model training with fewer labeled examples, reducing computational costs, and improving generalization to unseen cases [133]. Leveraging pre-trained weights also minimizes overfitting, which is a major concern when working with small datasets, making transfer learning a fundamental technique for modern computer-aided diagnosis.

In medical imaging, TL is broadly categorized as follows:

- *Feature-extractor based TL*: A convolutional neural network (CNN) pretrained on a large dataset (e. g., ImageNet) is used as a fixed feature generator; only task-specific classifier layers (e. g., fully connected layers) are retrained on the medical dataset [133].
- *Fine-tuning (model-based TL)*: Pretrained weights—fully or partially—are retrained on target medical data. Strategies range from retraining only classifier layers to unfreezing convolutional blocks or retraining the entire network [133].
- *Hybrid TL*: A two-stage approach beginning with feature extraction, followed by selective fine-tuning of deeper layers if performance is inadequate [133].

Transfer learning can also be categorized based on the supervision and domain relationship:

- *Inductive TL*: The source and target tasks differ, but labeled target data are available and used in fine-tuning.
- *Transductive TL (Domain adaptation)*: The task remains the same but domain distributions differ (e. g., different scanners or populations).
- *Self-supervised TL*: Models are pretrained via proxy tasks (e. g., contrastive, generative, or predictive) on unlabeled medical images and then fine-tuned using limited labeled data [261, 307].

11.1.2 Overcoming ML and DL Limitations with transfer learning

Deep learning modelsparticularly convolutional neural networks (CNNs)—have achieved remarkable success in medical image analysis, setting state-of-the-art benchmarks in tasks such as disease classification, lesion detection, and organ segmentation. However, despite their effectiveness, these models encounter several limitations, including the need for large annotated datasets, high computational costs, and susceptibility to overfitting when trained on small datasets. Transfer learning has emerged as a powerful solution to address these challenges by reusing knowledge from pretrained models and adapting it to specialised medical imaging tasks.

- *Limited data availability*: Deep models usually learn best from large, well-curated label sets. In clinical imaging, such corpora are hard to assemble: Expert annotation is expensive, time-consuming, and many diseases (e. g., rare cancer subtypes) are underrepresented. A practical workaround is to start from networks pre-trained on broad natural-image corpora and adapt them to the target task. Because these

models already encode generic visual structure, useful performance can often be reached with far fewer labels. For example, fine-tuning a pretrained CNN for colonoscopy polyp detection outperformed training the same architecture from scratch [286], illustrating the value of transfer learning when data are scarce.
- *Overfitting risks*: Small datasets invite memorization. Initializing from weights learned on diverse images provides a strong inductive bias: Early layers contribute stable edge/texture detectors, while later layers are nudged toward the clinical objective during fine-tuning. In practice, this reduces variance and improves generalization compared with randomly initialized training on the same data.
- *Computational efficiency*: Training large CNNs (or ViTs) end-to-end is compute-intensive and can be impractical for many labs. Fine-tuning pretrained backbones is markedly cheaper and faster, lowering the hardware and time budget needed to reach a useful baseline [133].

A concrete illustration comes from chest radiography: Systems initialized from pretrained backbones have achieved strong pneumonia detection with only a few hundred labeled images, an outcome that would be difficult to match with scratch training under the same constraints. More broadly, analyses of transfer in medical imaging report that pretraining improves both convergence and downstream accuracy in data-limited settings [222].

11.2 Transfer learning models

Transfer learning has revolutionized medical image analysis by leveraging pretrained CNN architectures that have proven effective across a range of tasks. The following models are among the most widely used due to their robust performance and versatility:
- *VGG16*: Developed by the Visual Geometry Group, VGG16 is a 16-layer CNN known for its simple yet deep architecture, which excels at extracting hierarchical features. It is frequently used for tasks like chest X-ray classification, where it can distinguish between normal and abnormal scans [259].
- *ResNet*:Residual Networks (ResNets), such as ResNet50 and ResNet101, introduced the concept of skip connections to effectively address the vanishing gradient problem. This innovation allows very deep neural networks to be trained without significant degradation in performance, enabling them to capture richer hierarchical features. In the medical imaging domain, ResNet architectures have been successfully applied to critical tasks, including tumor detection in CT scans and the classification of pulmonary nodules in lung images. For instance, [247] demonstrated that residual networks could achieve high accuracy in pulmonary nodule classification, underscoring their effectiveness in handling complex medical imaging challenges.
- *Inception*: The Inception-v3 model uses multiscale convolutional filters to capture diverse features, making it well-suited for complex tasks like detecting diabetic

retinopathy in retinal images, where subtle patterns are critical for accurate diagnosis.
- *DenseNet*: Dense convolutional networks connect each layer to every subsequent layer, promoting feature reuse and reducing the number of parameters. This efficiency makes DenseNet ideal for segmentation tasks, such as delineating tumors in MRI scans [222].

These architectures are commonly initialized with weights derived from pretraining on large-scale datasets such as ImageNet, which includes millions of natural images across thousands of categories. Such pretraining equips the network with a diverse set of low- and mid-level visual features—such as edges, textures, and shapes—that can be effectively transferred. Pretrained vision backbones are typically fine-tuned on domain-specific datasets to meet clinical objectives. In day-to-day practice, this means adapting a network on cohorts curated for tasks such as lung nodule detection or breast-lesion classification. Empirical evidence supports the approach: On relatively small CT datasets, fine-tuning a ResNet substantially improved nodule detection compared with training the same architecture from scratch, highlighting the practical benefits of transfer in this setting [259].

To shrink the gap between natural-image pretraining and medical-image target domains, recent work has explored two complementary directions. First, self-supervised pretraining enables models to learn useful representations directly from large unlabeled collections of medical scans, reducing dependence on costly expert labels. Second, multistage transfer learning proceeds in steps—initial pretraining on a broad corpus such as ImageNet, intermediate adaptation on a larger, mixed medical dataset, and a final fine-tune on the specific clinical task.

11.2.1 Common architectures and transfer learning techniques in medical imaging

As a practical workhorse in medical image analysis, transfer learning repurposes pretrained models for computer-aided diagnosis, tailoring them to the specifics of X-rays, MRIs, and CT scans. Various types of transfer learning are applied in this field, including traditional fine-tuning, feature extraction, domain adaptation, multistage transfer learning, and self-supervised learning. Each of these approaches offers a means to address challenges such as limited labeled data and the specialized requirements of medical imaging [205]. A common strategy is traditional fine-tuning, where the weights of pretrained convolutional neural networks (CNNs)—including architectures such as VGG16, ResNet, Inception-v3, and DenseNet—are adapted from large-scale datasets like ImageNet to domain-specific medical tasks [133]. For instance, there are reports that fine-tuned ResNet models attain classification accuracies exceeding 90 % in the detection of pneumonia from chest X-rays [259].

Feature extraction approaches, where early layers of a pretrained network are frozen and their learned representations are used as input to a new classifier, are especially useful when computational resources or labelled data are limited. For instance, in a comparison on the LUNA16 lung nodule detection challenge, a VGG16 feature-extraction model (freezing some layers) achieved a sensitivity of 90.78 % while maintaining high specificity [20]. Domain-adaptation models focus on aligning features between natural images and medical images, reducing the domain gap to improve performance in tasks like tumor segmentation from MRIs [222]. Multistage transfer learning sequentially fine-tunes models on intermediate medical datasets before targeting specific diagnostic tasks, achieving up to 98 % accuracy in hemorrhage detection in head CT scans [183]. Self-supervised learning, which generates representations without labeled data, enhances model robustness for tasks like diabetic retinopathy diagnosis, where DenseNet has reported an AUC-ROC of 0.95 [183].

11.2.2 Practical insights and limitations

Fine-tuning deep convolutional neural networks (CNNs), such as ResNet, has been shown to improve diagnostic accuracy in medical imaging. For instance, Tajbakhsh et al. [286] demonstrated substantial performance gains in colorectal polyp detection when fine-tuning pre-trained models compared to training from scratch. In mammography, adapting Inception-v3 has yielded reported sensitivities close to 92 % for breast-cancer detection. Performance is less uniform for rare subtypes, which remain harder to classify reliably [259].

Transfer learning is also useful when resources are tight. For tuberculosis screening on chest X-rays, a VGG16 backbone used only as a feature extractor achieved high specificity, while keeping compute and labeling needs modest [133]. When scanners or patient cohorts differ, representation alignment via domain adaptation has improved results in brain-tumor classification [222]. With very limited labels, multistage pipelines–pretraining on a broad source, adapting on a larger medical corpus, and then fine-tuning on the target task—have reduced overfitting and improved generalization for brain-tumor segmentation [183].

Dermatology offers a similar story. Because expert-annotated lesion datasets are small, self-supervised learning on unlabeled images has become appealing; models pre-trained this way have reached competitive accuracy using far fewer labels [133]. Together, these examples show how TL and SSL can adapt across a range of diagnostic settings. Caveats still apply. Outcomes depend strongly on dataset size and quality, as well as task complexity. Class imbalance, in particular, can inflate accuracy; careful augmentation and class-aware evaluation are needed to recover sensitivity for underrepresented findings [259]. Moreover, when large, well-curated labeled datasets are available, the relative advantage of transfer learning narrows [222].

Overall, the literature points to substantial gains in efficiency and accuracy from transfer learning, provided models and pretraining strategies are matched to the clinical context. In practice, this means choosing backbones, adaptation schemes, and validation protocols with the target use case in mind [183].

11.3 Performance analysis of transfer learning models in disease diagnosis

11.3.1 Evaluating transfer learning in medical image analysis

Assessing the performance of transfer learning (TL) models in medical image analysis requires a nuanced approach, as clinical applications differ in scope, modality, and diagnostic requirements. Relying on a single metric is often insufficient, as the clinical impact of false positives and false negatives varies across contexts. Therefore, the following multiple evaluation metrics are employed:

- *Accuracy:* Represents the overall proportion of correctly classified samples. While commonly reported, it can be misleading in highly imbalanced datasets. Accuracy may be inflated by the majority class, masking poor performance on minority classes. Additional metrics are necessary for a more accurate assessment [310].
- *Sensitivity (Recall):* The fraction of true positives that the model successfully identifies. High sensitivity is crucial in clinical settings, as missed cases (e. g., undetected tumors or fractures) can have serious consequences [310].
- *Specificity:* The fraction of true negatives correctly ruled out. Strong specificity limits false alarms, avoiding unnecessary tests, treatments, and patient anxiety [310].
- *Precision:* Among the cases flagged as positive, the fraction that are truly positive. High precision ensures alerts are credible and reduces unnecessary interventions in medical imaging [310].
- *F1 Score:* The harmonic mean of precision and recall. It balances missed detections and false alarms and is particularly useful when classes are imbalanced [310].
- *Area Under the ROC Curve (AUC):* A threshold-independent metric that summarizes how well the model separates positives from negatives across operating points. Larger AUC indicates stronger discrimination [310].
- *Dice Similarity Coefficient (DSC):* Measures overlap between predicted and reference regions for segmentation tasks. Higher DSC indicates more accurate boundary delineation [322].
- *Intersection over Union (IoU):* Measures the ratio of intersection to union between predicted and reference regions, commonly used for detection and segmentation quality assessment [304].
- *Mean Average Precision (mAP):* Summarizes precision–recall performance across multiple classes, commonly used for multiclass detection tasks [305].

Transfer learning-based systems frequently report strong discrimination, with AUC and sensitivity values often above 0.90 for disease detection. For segmentation, Dice and IoU scores commonly exceed 0.80, demonstrating that pretrained encoders adapt well to spatially detailed predictions. However, performance may degrade on datasets from different sites or scanners due to protocol differences, population shifts, and prevalence changes, emphasizing the importance of multimetric reporting and cross-dataset validation. Additionally, practical indicators such as speed of convergence during fine-tuning, performance under few-label conditions, and stability across institutions or cohorts are tracked to evaluate real-world applicability and generalizability of TL models.

TL-based systems in medical imaging frequently report strong discrimination, with AUC and sensitivity values often around or above 0.90 for disease-detection tasks. For segmentation, Dice and IoU scores commonly exceed 0.80, indicating that pretrained encoders adapt well to spatially detailed predictions. However, performance can degrade when models are evaluated on datasets from different sites or scanners, reflecting protocol differences, population shifts, and prevalence changes. This underscores the importance of multimetric reporting and explicit cross-dataset validation for clinical robustness. Beyond headline metrics, researchers also track practical signs of readiness: speed of convergence during fine-tuning, behavior under few-label conditions, and stability when shifting across institutions or cohorts. These aspects are crucial for assessing the real-world applicability and generalizability of TL models in medical settings.

Comparative studies generally favor transfer over training from scratch when labels are limited. For example, across several medical imaging tasks, fine-tuned CNNs outperformed scratch-trained counterparts, with the advantage most pronounced in small-data regimes [286]. In chest radiography, pretraining on natural images did not always deliver large gains with abundant labels, but it improved both efficiency and accuracy when annotations were scarce [222]. For 3D segmentation, initializing from Med3D weights boosted downstream Dice scores across multiple targets relative to ImageNet-based and scratch baselines [50]. More recently, self-supervised pretraining has added another route to transferable features: Models trained on unlabeled images have shown consistent improvements in low-label settings and under domain shift. Azizi et al. (2021) [22] presented a large-scale contrastive learning method (MICLe) and showed substantial improvements in top-1 accuracy and AUC across multiple medical image classification benchmarks.

In summary, general trends indicate that

- TL models provide consistent improvements in both classification and segmentation tasks when labeled data is limited;
- Pretrained models often converge faster and more stably than randomly initialized models;
- Domain-specific pretraining (e. g., on medical data) yields better results than using models pretrained on natural images;
- The benefits of TL reduce as the volume and quality of labeled medical data increase.

11.3.2 Automated diagnosis of pneumonia and COVID-19 from chest radiographs

CheXNet (DenseNet121) trained in ChestXray14 and fine-tuned for pneumonia detection achieved AUC greater than 0.90 and exceeded the average radiologist performance in the F1 score in held-out tests [225].

Chowdhury et al. evaluated AlexNet, ResNet-18, DenseNet-201, and SqueezeNet, fine-tuned end-to-end in 5,247 pediatric chest radiographs. They achieved up to 98 % precision (normal vs. pneumonia) and around 95 % precision for the classification of bacterial vs. viral pneumonia [54]. Another study using a public dataset of approximately 5,868 chest radiographs reported a peak accuracy of 98.43 % and an AUC exceeding 0.99 using ResNet-18 and DenseNet.

In COVID-19 classification tasks, several studies have applied transfer learning using ResNet-based frameworks. Abbas et al. [3] introduced DeTraC, a ResNet-18-based model, achieving 95. 12 % precision and 97. 91 % sensitivity to distinguish COVID-19 from normal and SARS cases. Similarly, Narin et al. [193] evaluated multiple ResNet architectures, finding that ResNet-50 achieved up to 96. 1 % precision and up to 99. 7 % in select datasets, with a sensitivity of around 98 % and an AUC close to 0.99 [193].

11.3.3 Segmentation and 3D imaging: Med3D and beyond

Accurate segmentation underpins many clinical workflows, such as defining tumor margins, estimating organ volumes, and planning interventions. Recent progress has come from deep models tailored to volumetric data, particularly 3D CNNs. Transfer learning (TL) adds a further boost: Starting from weights learned on related imaging corpora shortens training and often yields better baselines than learning everything from scratch.

A notable example is Med3D by Chen et al. [50]. The framework pretrains 3D CNN backbones on 3DSeg-8, a composite collection that spans eight public 3D segmentation datasets across organs and modalities. Those pretrained weights transfer cleanly to downstream targets such as liver-tumor segmentation (LiTS) and lung-nodule analysis (LIDC-IDRI), providing faster convergence than scratch training and outperforming pretraining in nonmedical sources such as Kinetics. In LiTS, reported Dice scores are around 94.6 %, reflecting the gains from domain-relevant pretraining [50].

Complementary architectural advances include UNet++, a nested U-Net design that strengthens multiscale feature aggregation and has been widely adopted in medical segmentation pipelines [354]. Though originally designed for 2D data, variants of UNet++ have been successfully adapted for 3D imaging and enhanced with pretrained encoders (e. g., ResNet-based backbones trained on large natural image datasets), significantly improving generalization. Isensee et al. proposed nnU-Net, a self-adapting framework that automatically configures itself for any given biomedical segmentation task [115].

Although nnU-Net is trained from scratch in its default configuration, variants incorporating transfer learning, particularly by initializing encoder weights from natural image models, have shown performance gains, especially when labeled data is scarce.

In the domain of brain tumor segmentation, Myronenko (2018) [187] introduced a 3D VAE-UNet architecture that incorporates a variational autoencoder (VAE) branch to learn robust latent representations. Although the original model was not pretrained, later adaptations that integrated ImageNet-pre-trained encoders demonstrated improved performance on the BraTS dataset. Similarly, Mondal et al. [180] investigated the use of pretrained 2D CNNs, such as VGG-16, as slice-wise feature extractors within hybrid 2D/3D segmentation frameworks. Their experiments on 3D cardiac magnetic resonance imaging data revealed that transfer learning not only accelerated training but also improved segmentation accuracy, particularly under limited-data conditions.

In sum, transfer learning occupies a central role in 3D medical image segmentation by mitigating limited annotations, reducing optimization time, and improving robustness to intermodality differences.

11.3.4 Self-supervised pretraining: enhanced transferability

Self-supervised learning (SSL) has emerged as a credible way to improve transferability in medical imaging, especially when labels are scarce. Unlike conventional supervised pretraining, SSL learns from unlabeled data by posing proxy objectives, and the resulting features often carry over well across modalities and tasks.

A representative example is MICLe (Multi-Instance Contrastive Learning) by Azizi et al. [22]. The recipe combines SSL pretraining (initialized from ImageNet) with task-specific fine-tuning on medical datasets such as chest radiographs and dermatology images. Reported gains include roughly a 6.7 % increase in top-1 accuracy and about a 1.1 % increase in mean AUC over strongly supervised baselines.

Beyond MICLe, several SSL families, including SimCLR, SwAV, and DINO, have been explored for clinical use. When these methods were trained on mixes of natural and medical images, investigators observed improvements as large as 14.8 % in Cohen's kappa and around 5.4 % in AUC for tumor detection, with additional gains for diabetic retinopathy and classification of chest pathology [300].

The evidence base is not limited to isolated case studies. A recent synthesis in X-ray, CT, MRI, and ultrasound concluded that SSL-pretrained models tend to outperform fully supervised pretraining and training from scratch, particularly under domain shift and low-label conditions [310].

In particular, large-scale nonmedical SSL can also be competitive. Using DINOv2-style pretraining, Tayebi et al. reported performance that surpassed ImageNet-supervised models and, in some settings, even task-specific supervised training [291]. Interestingly, their method outperformed models trained on more than 800,000 chest radiographs for more than 20 distinct findings. In the context of screening for diabetic retinopathy, Ouyang et al. (2022) [201] introduced SimCLR-DR, a contrastive learning framework that uses unlabeled images of the retinal fundus. Their results showed that SimCLR-DR outperformed ImageNet-pretrained baselines, especially when labeled data was limited.

For chest X-ray interpretation, Sun et al. (2023) [277] proposed a SimCLR-based SSL approach that significantly improved the sensitivity and generalization of rib fracture detection, with particular effectiveness in low-data scenarios. Lastly, Zeng et al. (2024) [338] provided a comprehensive survey of SSL methods in medical imaging, highlighting how contrastive SSL techniques have driven advances in both segmentation and classification in CT, MRI, ultrasound, histopathology, and X-ray applications.

11.4 Practical insights and limitations

Transfer learning, whether through supervised or self-supervised pretraining, has emerged as a foundational strategy for medical AI. Its practical benefits span across various imaging modalities, clinical tasks, and data-availability settings.

- *Improved accuracy and robustness*: Models pretrained using SSL or supervised learning in large-scale datasets frequently outperform those trained from scratch and sometimes even surpass the performance of experts in diagnostic and classification tasks [22, 291].
- *Faster convergence and greater training stability*: Pretrained models, especially those initialized with domain-specific features or contrastive objectives, tend to require fewer epochs and exhibit more stable learning dynamics during fine-tuning [201, 300].
- *Label efficiency*: SSL methods have proven especially effective in low-label regimes, achieving high accuracy with as few as 100 labeled samples per class. This is particularly valuable in rare disease detection, pediatric imaging, and early-stage screening tasks [277, 338].
- *Cross-modality generalization*: Pretraining on one modality (e. g., X-rays) can still benefit downstream tasks in other modalities (e. g., CT or MRI), depending on the learned representation space and fine-tuning strategy [310].
- *Limitations and challenges*: SSL methods typically require very large unlabeled datasets (millions of samples) and access to high-performance computing infrastructure. Moreover, the pretraining objectives (e. g., contrastive vs. masked modeling) must be carefully matched to the target task. Improper transfer can result in poor performance or negative transfer, especially if the pretraining domain diverges significantly from the downstream clinical task [291, 300].
- *Need for architectural alignment*: Not all pre-trained backbones transfer equally well – architectures such as Vision Transformers (ViTs) may require different augmentation strategies or fine-tuning protocols compared to CNN-based models [201].

11.5 Future directions and conclusions

Transfer learning, both supervised and self-supervised, has emerged as a powerful paradigm in medical image analysis. However, real-world deployment poses multiple challenges, which require continued innovation in algorithmic robustness, clinical relevance, and ethical compliance, such as the following:

1. *Domain shift*: Medical images (e. g., CT, MRI, X-ray) often differ in modality, scale, and structure from natural images like those in ImageNet, resulting in a domain gap that weakens transferability. Domain-adaptation methods and modality-sensitive pretraining strategies are being explored to bridge this gap, although performance remains inconsistent across tasks [222].
2. *Model selection and fine-tuning*: There is no universal strategy for selecting the best model or tuning its parameters. The effectiveness of transfer learning often depends on factors such as model architecture (e. g., CNN vs. ViT), task complexity, dataset size, and label distribution. Finding the right configuration, such as freezing layers, learning rate scheduling, or hybrid training, requires empirical tuning and domain experience [133].
3. *Computational constraints*: Although transfer learning reduces the training burden, models like DenseNet, ResNet, or Vision Transformers can still be resource-intensive during fine-tuning. Limited access to high-performance computing resources, especially in smaller institutions or clinical settings, restricts scalability and reproducibility.
4. *Model interpretability*: The lack of transparency in deep learning models remains a significant barrier to clinical integration. Tools such as Grad-CAM, saliency maps, and attention-based visualizations aim to address this issue, but often do not provide clinically actionable insights or undergo rigorous validation [259].
5. *Data privacy and regulation*: Regulatory policies like HIPAA and GDPR limit access to labeled medical data, restricting training and evaluation pipelines. Techniques like federated learning and secure multiparty computation enable decentralized learning without sharing raw data, but add layers of complexity, and are not yet widely adopted in clinical settings [183].

Bibliography

[1] Seq-diff-net: Sequential differencing network for melanoma evolution. *Pattern Recognition Letters*, 156:34–42, 2022.

[2] Ckdnet: Contextual knowledge diffusion network for skin lesion analysis. *IEEE Access*, 11:78912–78925, 2023.

[3] A. Abbas, M. M. Abdelsamea, and M. M. Gaber. Classification of covid-19 in chest X-ray images using detrac deep convolutional neural network. *IEEE Access*, 8:74027–74035, 2020.

[4] I. Aboussaleh, J. Riffi, A. M. Mahraz, and H. Tairi. Inception-UDet: An improved U-Net architecture for brain tumor segmentation. *Annals of Data Science*, 11(3):831–853, 2024.

[5] M. D. Abràmoff, P. T. Lavin, M. Birch, J. N. Shah, and J. C. Folk. Pivotal trial of an autonomous ai-based diagnostic system for detection of diabetic retinopathy in primary care offices. *npj Digital Medicine*, 1(1):39, 2018.

[6] K. Aderghal, A. Khvostikov, A. Krylov, J. Benois-Pineau, K. Afdel, and G. Catheline. Classification of alzheimer disease on imaging modalities with deep cnns using cross-modal transfer learning. In *Proc. IEEE 31st Int. Symp. Computer-Based Med. Syst. (CBMS)*, pages 345–350, 2018.

[7] K. D. Ahmed and R. Hawezi. Detection of bone fracture based on machine learning techniques. *Measurement: Sensors*, 27:100723, 2023.

[8] K. H. Ali. Vit-bt: Improving mri brain tumor classification using vision transformer with transfer learning. Preprint, SSRN, 2023.

[9] M. Z. Alom, C. Yakopcic, M. S. Nasrin, T. M. Taha, and V. K. Asari. Breast cancer classification from histopathological images with inception recurrent residual convolutional neural network. *Journal of Digital Imaging*, 32(4):605–617, 2019.

[10] O. Alpar, R. Dolezal, P. Ryska, and O. Krejcar. Low-contrast lesion segmentation in advanced mri experiments by time-domain ricker-type wavelets and fuzzy 2-means. *Applied Intelligence*, 1–22, 2022.

[11] S. Alqazzaz, X. Sun, L. D. M. Nokes, H. Yang, Y. Yang, R. Xu, Y. Zhang, and X. Yang. Combined features in region of interest for brain tumor segmentation. *Journal of Digital Imaging*, 1–9, 2022.

[12] S. M. Alqhtani, T. A. Soomro, A. A. Shah, A. A. Memon, M. Irfan, S. Rahman, and L. A. B. Eljak. Improved brain tumor segmentation and classification in brain MRI with FCM–SVM: a diagnostic approach. *IEEE Access*, 12:61312–61335, 2024.

[13] American Cancer Society. *Cancer facts & figures 2022*. American Cancer Society, Atlanta, GA, USA, 2022.

[14] D. Amiel, J. B. Kleiner, and W. H. Akeson. Anterior cruciate ligament injury and repair: clinical and experimental perspectives. *Journal of Orthopaedic Research*, 7(4):472–475, 1989.

[15] Medscape Editorial Contributors. Anterior cruciate ligament injury: Background, functional anatomy, sport-specific biomechanics. https://emedicine.medscape.com/article/1252414-clinical. Accessed: 2025-07-21.

[16] R. Arora, B. Raman, K. Nayyar, and R. Awasthi. Automated skin lesion segmentation using attention-based deep convolutional neural network. *Biomedical Signal Processing and Control*, 65:102358, 2021.

[17] M. Arsalan, M. Owais, T. Mahmood, S. W. Cho, and K. R. Park. Aiding the diagnosis of diabetic and hypertensive retinopathy using artificial intelligence-based semantic segmentation. *Journal of Clinical Medicine*, 8(9):1446, 2019.

[18] S. Arumugam, S. Paulraj, and N. P. Selvaraj. Brain mr image tumor detection and classification using neuro fuzzy with binary cuckoo search technique. *International Journal of Imaging Systems and Technology*, 31(3):1185–1196, 2021.

[19] A. A. Asiri, T. A. Soomro, A. A. Shah, G. Pogrebna, M. Irfan, and S. Alqahtani. Optimized brain tumor detection: a dual-module approach for MRI image enhancement and tumor classification. *IEEE Access*, 12:42868–42887, 2024.

https://doi.org/10.1515/9783111389059-012

[20] J. Gao, Q. Jiang, B. Zhou, and D. Chen. Lung nodule detection using convolutional neural networks with transfer learning on CT images. *Combinatorial Chemistry & High Throughput Screening*, 24(6):814–824, 2021.
[21] G. Ayana, K. Dese, Y. Dereje, Y. Kebede, H. Barki, D. Amdissa, N. Husen, F. Mulugeta, B. Habtamu, and S.-W. Choe. Vision-transformer-based transfer learning for mammogram classification. *Diagnostics*, 13(2):178, 2023.
[22] S. Azizi, B. Mustafa, F. Ryan, Z. Beaver, J. Freyberg, J. Deaton, A. Loh, A. Karthikesalingam, S. Kornblith, T. Chen, et al. Big self-supervised models advance medical image classification. In *Proceedings of the IEEE/CVF International Conference on Computer Vision (ICCV)*, pages 3478–3488, 2021.
[23] Bach: Breast cancer histology dataset. https://data.niaid.nih.gov/resources?id=zenodo_3632034. Accessed 2025.
[24] U. Baid, S. U. Rane, S. Talbar, S. Bakas, et al. The rsna-asnr-miccai brats 2021 benchmark on brain tumor segmentation and radiogenomic classification. *arXiv preprint arXiv:2107.02314*, 2021.
[25] S. Bakas, H. Akbari, A. Sotiras, M. Bilello, M. Rozycki, J. Kirby, J. Freymann, K. Farahani, and C. Davatzikos. Advancing the cancer genome atlas glioma mri collections with expert segmentation labels and radiomic features. *Scientific Data*, 4:170117, 2017.
[26] S. Bakas, M. Reyes, A. Jakab, et al. Identifying the best machine learning algorithms for brain tumor segmentation, progression assessment, and overall survival prediction in the BRATS challenge. *arXiv preprint arXiv:1811.02629*, 2018.
[27] T. Balamurugan and E. Gnanamanoharan. Brain tumor segmentation and classification using hybrid deep CNN with LuNetClassifier. *Neural Computing & Applications*, 35(6):4739–4753, 2023.
[28] L. Ballerini, R. B. Fisher, B. Aldridge, and J. Rees. A color and symmetry based feature set for the automatic classification of dermoscopy images. In *ICIP*, pages 5437–5440, 2013.
[29] T. Baltrušaitis, C. Ahuja, and L.-P. Morency. Multimodal machine learning: A survey and taxonomy. *IEEE TPAMI*, 41(2):423–443, 2019.
[30] Y. M. Baltruschat, H. Nickisch, M. Grass, T. Knopp, and A. Saalbach. Comparison of deep learning approaches for multi-label chest x-ray classification. *Scientific Reports*, 9(1):6381, 2019.
[31] O. Bandyopadhyay, A. Biswas, and B. Bhattacharya. Long-bone fracture detection in digital x-ray images based on digital-geometric techniques. *Computer Methods and Programs in Biomedicine*, 123:2–14, 2016.
[32] Z. Barzegar and M. Jamzad. Wlfs: Weighted label fusion learning framework for glioma tumor segmentation in brain mri. *Biomedical Signal Processing and Control*, 68:102617, 2021.
[33] C. M. A. K. Z. Basha, T. M. Padmaja, and G. N. Balaji. An effective and reliable computer automated technique for bone fracture detection. *EAI Endorsed Transactions on Pervasive Health and Technology*, 5(18):e2, 2019.
[34] N. Bayramoglu, J. Kannala, and J. Heikkilä. Deep learning for magnification independent breast cancer histopathology image classification. In *Pattern Recognition (ICPR), 23rd International Conference on*, pages 2440–2445. IEEE, 2016.
[35] L. A. Beckett, M. C. Donohue, C. Wang, P. Aisen, D. J. Harvey, and N. Saito, Alzheimer's Disease Neuroimaging Initiative, et al. The alzheimer's disease neuroimaging initiative phase 2: increasing the length, breadth, and depth of our understanding. *Alzheimer's & Dementia*, 11(7):823–831, 2015.
[36] M. Behzadpour, E. Azizi, K. Wu, and B. L. Ortiz. Enhancing brain tumor segmentation using channel attention and transfer learning. *arXiv preprint arXiv:2501.11196*, 2025.
[37] J. Bertels, T. Eelbode, M. Berman, D. Vandermeulen, F. Maes, R. Bisschops, and M. B. Blaschko. Optimizing the dice score and jaccard index for medical image segmentation: Theory and practice. In *Medical Image Computing and Computer Assisted Intervention–MICCAI 2019: 22nd International Conference, Shenzhen, China, October 13–17, 2019, Proceedings, Part II 22*, pages 92–100. Springer, 2019.
[38] B. D. Beynnon and B. C. Fleming. The biomechanics of anterior cruciate ligament injury in the animal model. *Operative Techniques in Sports Medicine*, 13(3):156–162, 2005.

[39] U. A. Bhatti, J. Liu, M. Huang, and Y. Zhang. FF-UNet: Feature fusion based deep learning-powered enhanced framework for accurate brain tumor segmentation in MRI images. *Image and Vision Computing*, 105635, 2025.

[40] X. Bian, X. Luo, C. Wang, W. Liu, and X. Lin. Optic disc and optic cup segmentation based on anatomy guided cascade network. *Computer Methods and Programs in Biomedicine*, 197:105717, 2020.

[41] Y. Bian, C. Si, and L. Wang. Diabetic retinopathy lesion segmentation method based on multi-scale attention and lesion perception. *Algorithms*, 17(4):161, 2024.

[42] K. Blennow, M. J. de Leon, and H. Zetterberg. Alzheimer's disease. *The Lancet*, 368(9533):387–403, 2006.

[43] L. Bloch and C. M. Friedrich. Systematic comparison of 3d deep learning and classical machine learning explanations for alzheimer's disease detection. *Computers in Biology and Medicine*, 170:108029, 2024.

[44] H. Bolhasani, E. Amjadi, M. Tabatabaeian, and S. J. Jassbi. A histopathological image dataset for grading breast invasive ductal carcinomas. *Informatics in Medicine Unlocked*, 19:100341, 2020.

[45] Breakhis: Breast cancer histopathological database. https://web.inf.ufpr.br/vri/databases/breast-cancer-histopathological-database-breakhis/. Accessed 2025.

[46] M. E. Celebi, Q. Wen, S. Hwang, H. Iyatomi, and G. Schaefer. Lesion border detection in dermoscopy images using ensembles of thresholding methods. *Skin Research and Technology*, 19(1):e252–e258, 2013.

[47] P. D. Chang, T. T. Wong, and M. J. Rasiej. Deep learning for detection of complete anterior cruciate ligament tear. *Journal of Digital Imaging*, 32:980–986, 2019.

[48] J. Chen, Y. Lu, Q. Yu, X. Luo, E. Adeli, Y. Wang, L. Lu, A. L. Yuille, and Y. Zhou. Transunet: Transformers make strong encoders for medical image segmentation. *arXiv preprint arXiv:2102.04306*, 2021.

[49] L. C. Chen, Y. Zhu, G. Papandreou, F. Schroff, and H. Adam. Encoder-decoder with atrous separable convolution for semantic image segmentation. In *European Conference on Computer Vision (ECCV)*, pages 801–818, 2018.

[50] S. Chen, K. Ma, and Y. Zheng. Med3d: Transfer learning for 3d medical image analysis. *arXiv preprint arXiv:1904.00625*, 2019.

[51] Q. Cheng, H. Lin, J. Zhao, X. Lu, and Q. Wang. Application of machine learning-based multi-sequence mri radiomics in diagnosing anterior cruciate ligament tears. *Journal of Orthopaedic Surgery and Research*, 19(1):99, 2024.

[52] D. Chicco and G. Jurman. The advantages of the matthews correlation coefficient (mcc) over f1 score and accuracy in binary classification evaluation. *BMC Genomics*, 21(1):6, 2020.

[53] H. Chougrad, H. Zouaki, and O. Alheyane. Multi-label transfer learning for the early diagnosis of breast cancer. *Neurocomputing*, 392:168–180, 2020.

[54] T. R. Chowdhury, M. Reaz, ..., S. Kashem, and A. Khandakar, Transfer learning with deep convolutional neural network (cnn) for pneumonia detection using chest x-ray. *arXiv preprint arXiv:2004.06578*, 2020.

[55] N. Christine. Your diabetic patients: look them in the eyes. which ones will lose their sight? 2015. https://www.eyepacs.com/diabeticretinopathy/.

[56] U. Chutia, A. S. Tewari, and J. P. Singh. Collapsed lung disease classification by coupling denoising algorithms and deep learning techniques. *Network Modeling Analysis in Health Informatics and Bioinformatics*, 13(1):1, 2023.

[57] N. Cinar, A. Ozcan, and M. Kaya. A hybrid DenseNet121-UNet model for brain tumor segmentation from MR images. *Biomedical Signal Processing and Control*, 76:103647, 2022.

[58] N. C. Codella, D. Gutman, M. E. Celebi, B. Helba, M. A. Marchetti, S. W. Dusza, A. Kalloo, K. Liopyris, N. Mishra, H. Kittler, et al. Skin lesion analysis toward melanoma detection: A challenge at the 2017 international symposium on biomedical imaging (isbi), hosted by the international skin imaging collaboration (isic). In *Proceedings of the IEEE 15th International Symposium on Biomedical Imaging (ISBI)*, pages 168–172, 2018.

[59] D. L. Collins, A. P. Zijdenbos, V. Kollokian, J. G. Sled, N. J. Kabani, C. J. Holmes, and A. C. Evans. Design and construction of a realistic digital brain phantom. *IEEE Transactions on Medical Imaging*, 17(3):463–468, 1998.

[60] C. Cortes and V. Vapnik. Support-vector networks. *Machine Learning*, 20(3):273–297, 1995.

[61] D. Dablain, B. Krawczyk, and N. V. Chawla. Deepsmote: Fusing deep learning and smote for imbalanced data. *IEEE Transactions on Neural Networks and Learning Systems*, 2022.

[62] C. Dai, S. Wang, Y. Mo, E. Angelini, Y. Guo, and W. Bai. Suggestive annotation of brain mr images with gradient-guided sampling. *Medical Image Analysis*, 102373, 2022.

[63] L. Dai, B. Sheng, T. Chen, Q. Wu, R. Liu, C. Cai, ..., and W. Jia. A deep learning system for predicting time to progression of diabetic retinopathy. *Nature Medicine*, 30(2):584–594, 2024.

[64] L. Dai, L. Wu, H. Li, J. Cai, X. Wang, and W. Lin. A deep learning system for detecting diabetic retinopathy across the disease spectrum. *Nature Communications*, 12:1622, 2021.

[65] É. Decencière, X. Zhang, G. Cazuguel, D. Lesage, B. Cochener, C. Trone, P. Gain, I. Leclercq, M. Lamard, G. Cléroux, and C. Roux. Feedback on a publicly distributed database: the messidor database. *Image Analysis & Stereology*, 33(3):231–234, 2014.

[66] M. Degirmenci, Y. K. Yuce, M. Perc, and Y. Isler. Statistically significant features improve binary and multiple motor imagery task predictions from eegs. *Frontiers in Human Neuroscience*, 17, 2023.

[67] J. Deng, W. Dong, R. Socher, L.-J. Li, K. Li, and L. Fei-Fei. Imagenet: A large-scale hierarchical image database. In *Proceedings of the IEEE Conference on Computer Vision and Pattern Recognition (CVPR)*, pages 248–255, 2009.

[68] W. Deng, Q. Shi, M. Wang, B. Zheng, and N. Ning. Deep learning-based hcnn and crf-rrnn model for brain tumor segmentation. *IEEE Access*, 8:26665–26675, 2020.

[69] L. R. Dice. Measures of the amount of ecologic association between species. *Ecology*, 26(3):297–302, 1945.

[70] A. Dosovitskiy, L. Beyer, A. Kolesnikov, D. Weissenborn, X. Zhai, T. Unterthiner, M. Dehghani, M. Minderer, G. Heigold, S. Gelly, J. Uszkoreit, and N. Houlsby. An image is worth 16x16 words: Transformers for image recognition at scale. In *International Conference on Learning Representations*, 2021. arXiv:2010.11929.

[71] S. Dubey. Alzheimer's dataset (4 class of images). Images of MRI Segementation, 2019.

[72] A. Ebrahimi, S. Luo, and R. Chiong. Alzheimer's Disease Neuroimaging Initiative, et al. Deep sequence modelling for alzheimer's disease detection using mri. *Computers in Biology and Medicine*, 134:104537, 2021.

[73] N. A. El Joudi, M. Lazaar, F. Delmotte, H. Allaoui, and O. Mahboub. Adaptive transfer learning using segformer for imbalanced pixel-level medical image segmentation. *Signal, Image and Video Processing*, 19:617–629, 2025.

[74] C. W. Elston and I. O. Ellis. Pathological prognostic factors in breast cancer. i. the value of histological grade in breast cancer: experience from a large study with long-term follow-up. *Histopathology*, 19(5):403–410, 1991.

[75] A. Esteva, B. Kuprel, R. A. Novoa, J. Ko, S. M. Swetter, H. M. Blau, and S. Thrun. Dermatologist-level classification of skin cancer with deep neural networks. *Nature*, 542(7639):115–118, 2017.

[76] C. Feng and et al.. Deep learning framework for alzheimer's disease diagnosis via 3d-cnn and fsbi-lstm. *IEEE Access*, 7:63605–63618, 2019.

[77] A. Ferretti, P. Papandrea, F. Conteduca, and P. P. Mariani. Functional anatomy of the anterior cruciate ligament. *Knee Surgery, Sports Traumatology, Arthroscopy*, 1(2):79–83, 1991.

[78] G. Folego, M. Weiler, R. F. Casseb, R. Pires, and A. Rocha. Alzheimer's disease detection through whole-brain 3d-cnn mri. *Frontiers in Bioengineering and Biotechnology*, 8:534592, 2020.

[79] A. F. Fotenos, A. Z. Snyder, L. E. Girton, J. C. Morris, and R. L. Buckner. Normative estimates of cross-sectional and longitudinal brain volume decline in aging and ad. *Neurology*, 64(6):1032–1039, 2005.

[80] F. Fumero, S. Alayón, J. L. Sanchez, J. Sigut, and M. Gonzalez-Hernandez. Rim-one: An open retinal image database for optic nerve evaluation. In *2011 24th international symposium on computer-based medical systems (CBMS)*, pages 1–6. IEEE, 2011.

[81] W. Gale, L. Oakden-Rayner, G. Carneiro, A. P. Bradley, and L. J. Palmer. Detecting hip fractures with radiologist-level performance using deep neural networks. *arXiv preprint arXiv:1711.06504*, 2017.

[82] Y. Gao, M. Zhou, and D. N. Metaxas. Utnet: a hybrid transformer architecture for medical image segmentation. In *MICCAI*, 2021.

[83] S. J. S. Gardezi, A. Elazab, B. Lei, and T. Wang. Breast cancer detection and diagnosis using mammographic data: Systematic review. *Journal of Medical Internet Research*, 21:e13008, 2019.

[84] N. Gessert, M. Nielsen, M. Shaikh, R. Werner, and A. Schlaefer. Skin lesion classification using loss balancing and ensembles of multi-resolution efficientnets. In *ISIC*, 2019.

[85] T. M.Ghazal, S. Abbas, S. Munir, M. A. Khan, M. Ahmad, G. F. Issa, S. B. Zahra, M. A. Khan, and M. K. Hasan. Alzheimer disease detection empowered with transfer learning. *Computers, Materials & Continua*, 70(3), 2022.

[86] F. Ghazouani, P. Vera, and S. Ruan. Efficient brain tumor segmentation using swin transformer and enhanced local self-attention. *International Journal of Computer Assisted Radiology and Surgery*, 19(2):273–281, 2024.

[87] L. Gonog and Y. Zhou. A review: Generative adversarial networks. In *2019 14th IEEE Conference on Industrial Electronics and Applications (ICIEA)*, June 2019.

[88] R. Gu, L. Wang, and L. Zhang. De-net: A deep edge network with boundary information for automatic skin lesion segmentation. *Neurocomputing*, 468:71–84, 2022.

[89] V. Gulshan, L. Peng, M. Coram, M. C. Stumpe, et al. Development and validation of a deep learning algorithm for detection of diabetic retinopathy in retinal fundus photographs. *JAMA*, 316(22):2402–2410, 2016.

[90] P. Gupta, Z. Sinno, J. L. Glover, N. G. Paulter, and A. C. Bovik. Predicting detection performance on security x-ray images as a function of image quality. *IEEE Transactions on Image Processing*, 28(7):3328–3342, 2019.

[91] M. Hamdi, S. Bourouis, K. Rastislav, F. Mohmed, et al. Evaluation of neuro images for the diagnosis of alzheimer's disease using deep learning neural network. *Frontiers in Public Health*, 10:834032, 2022.

[92] Z. Han, B. Wei, Y. Zheng, Y. Yin, K. Li, and S. Li. Breast cancer multiclassification from histopathological images with structured deep learning model. *Scientific Reports*, 7(1):4172, 2017.

[93] M. Hany. Chest ct-scan images dataset. CT-Scan images with different types of chest cancer, 2020.

[94] R. Hao, K. Namdar, L. Liu, and F. Khalvati. A transfer learning–based active learning framework for brain tumor classification. *Frontiers in Artificial Intelligence*, 4:635766, 2021.

[95] F. Hardalaç, F. Uysal, O. Peker, M. Çiçeklidağ, T. Tolunay, N. Tokgöz, U. Kutbay, B. Demirciler, and F. Mert. Fracture detection in wrist x-ray images using deep learning-based object detection models. *Sensors*, 22(3):1285, 2022.

[96] M. K. Hasan, L. Dahal, P. N. Samarakoon, F. I. Tushar, and R. Martí. Dsnet: Automatic dermoscopic skin lesion segmentation. *Computers in Biology and Medicine*, 120:103738, 2020.

[97] A. Hatamizadeh, V. Nath, Y. Tang, D. Yang, H. R. Roth, and D. Xu. Swin unetr: Swin transformers for semantic segmentation of brain tumors in mri images. In *International MICCAI Brainlesion Workshop*, pages 272–284. Springer, 2021.

[98] K. He, X. Zhang, S. Ren, and J. Sun. Deep residual learning for image recognition. In *Proceedings of the IEEE Conference on Computer Vision and Pattern Recognition (CVPR)*, pages 770–778, 2016.

[99] Q. He, Q. Yang, and M. Xie. Hctnet: A hybrid cnn-transformer network for breast ultrasound image segmentation. *Computers in Biology and Medicine*, 155:106629, 2023.

[100] J. H. Ho, W. Z. Lung, and C. L. Seah. Anterior cruciate ligament segmentation: using morphological operations with active contour. In *2010 4th International Conference on Bioinformatics and Biomedical Engineering*, pages 1–4. IEEE, 2010.

[101] A. Holzinger, G. Langs, H. Denk, K. Zatloukal, and H. Müller. Causability and explainability of artificial intelligence in medicine. *Wiley Interdisciplinary Reviews: Data Mining and Knowledge Discovery*, 9(4):e1312, 2019.

[102] S. Hossain, A. Chakrabarty, T. R. Gadekallu, M. Alazab, and Md. J. Piran. Vision transformers, ensemble model, and transfer learning leveraging explainable ai for brain tumor detection and classification. *IEEE Access*, 9:113801–113811, 2021.

[103] S. Kumar HS and K. Karibasappa. An effective hybrid deep learning with adaptive search and rescue for brain tumor detection. *Multimedia Tools and Applications*, 81(13):17669–17701, 2022.

[104] C. I. Hsieh, K. Zheng, C. Lin, L. Mei, L. Lu, W. Li, and C. F. Kuo. Automated bone mineral density prediction and fracture risk assessment using plain radiographs via deep learning. *Nature Communications*, 12(1):5472, 2021.

[105] C. Huang, L.-O. Wahlund, O. Almkvist, D. Elehu, L. Svensson, T. Jonsson, B. Winblad, and P. Julin. Voxel- and voi-based analysis of spect cbf in relation to clinical and psychological heterogeneity of mild cognitive impairment. *NeuroImage*, 19(3):1137–1144, July 2003.

[106] G. Huang, Z. Liu, L. van der Maaten, and K. Q. Weinberger. Densely connected convolutional networks. In *Proceedings of the IEEE Conference on Computer Vision and Pattern Recognition (CVPR)*, pages 2261–2269, 2017.

[107] L. Huang, E. Zhu, L. Chen, Z. Wang, S. Chai, and B. Zhang. A transformer-based generative adversarial network for brain tumor segmentation. *Frontiers in Neuroscience*, 16:1054948, 2022.

[108] Z. Huang, Y. Liu, G. Song, and Y. Zhao. Gammanet: An intensity-invariance deep neural network for computer-aided brain tumor segmentation. *Optik*, 243:167441, 2021.

[109] Z. Huang, Y. Zhao, Y. Liu, and G. Song. GCAUNet: A group cross-channel attention residual UNet for slice based brain tumor segmentation. *Biomedical Signal Processing and Control*, 70:102958, 2021.

[110] D. P. Huttenlocher, G. A. Klanderman, and W. J. Rucklidge. Comparing images using the hausdorff distance. *IEEE Transactions on Pattern Analysis and Machine Intelligence*, 15(9):850–863, 1993.

[111] I. A. M. Ikhsan, A. Hussain, M. A. Zulkifley, N. M. Tahir, and A. Mustapha. An analysis of x-ray image enhancement methods for vertebral bone segmentation. In *2014 IEEE 10th international colloquium on signal processing and its applications*, pages 208–211. IEEE, 2014.

[112] A. E. Ilesanmi, U. Chaumrattanakul, and S. S. Makhanov. A method for segmentation of tumors in breast ultrasound images using the variant enhanced deep learning. *Biocybernetics and Biomedical Engineering*, 41(2):802–818, 2021.

[113] R. Imtiaz, T. M. Khan, S. S. Naqvi, M. Arsalan, and S. J. Nawaz. Screening of glaucoma disease from retinal vessel images using semantic segmentation. *Computers & Electrical Engineering*, 91:107036, 2021.

[114] E. L. Irede, O. R. Aworinde, O. K. Lekan, O. D. Amienghemhen, T. P. Okonkwo, A. P. Onivefu, and I. H. Ifijen. Medical imaging: a critical review on x-ray imaging for the detection of infection. *Biomedical Materials & Devices*, 1–45, 2024.

[115] F. Isensee, P. F. Jaeger, S. A. A. Kohl, J. Petersen, and K. H. Maier-Hein. nnu-net: a self-configuring method for deep learning-based biomedical image segmentation. *Nature Methods*, 18(2):203–211, 2021.

[116] F. Isensee, P. F. Jaeger, S. A. A. Kohl, J. Petersen, and K. H. Maier-Hein. nnu-net: a self-adapting framework for u-net-based medical image segmentation. *Nature Methods*, 18:203–211, 2021.

[117] S. Izadi, Z. Mirikharaji, J. Kawahara, and G. Hamarneh. Generative adversarial networks to segment skin lesions. In *2018 IEEE 15th International Symposium on Biomedical Imaging*, pages 881–884. IEEE, 2018.

[118] A. Janowczyk and A. Madabhushi. Deep learning for digital pathology image analysis: A comprehensive tutorial with selected use cases. *Journal of Pathology Informatics*, 7(1):29, 2016.

[119] Jboysen. Mri and alzheimer's. https://www.kaggle.com/datasets/jboysen/mri-and-alzheimers, 2020. Kaggle dataset, accessed 2025-09-11.

[120] J. Jeong. Eeg dynamics in patients with alzheimer's disease. *Clinical Neurophysiology*, 115(7):1490–1505, 2004.

[121] Q. Jia and H. Shu. Bitr-unet: a cnn-transformer combined network for mri brain tumor segmentation. In *Brainlesion: Glioma, Multiple Sclerosis, Stroke and Traumatic Brain Injuries*, Lecture Notes in Computer Science, pages 3–14. Springer, 2021.

[122] H. Jiang, J. Xu, R. Shi, K. Yang, D. Zhang, M. Gao, H. Ma, and W. Qian. A multi-label deep learning model with interpretable grad-cam for diabetic retinopathy classification. In *2020 42nd Annual International Conference of the IEEE Engineering in Medicine & Biology Society (EMBC)*, pages 1560–1563, 2020.

[123] M. Jiang, F. Zhai, and J. Kong. A novel deep learning model DDU-Net using edge features to enhance brain tumor segmentation on MR images. *Artificial Intelligence in Medicine*, 121:102180, 2021.

[124] Q. Jin, H. Cui, C. Sun, Z. Meng, and R. Su. Cascade knowledge diffusion network for skin lesion diagnosis and segmentation. *Applied Soft Computing*, 99:106881, 2021.

[125] D. Joshi and T. P. Singh. A survey of fracture detection techniques in bone x-ray images. *Artificial Intelligence Review*, 53(1):447–507, 2020.

[126] G. A. Kaissis, M. R. Makowski, D. Rückert, and R. F. Braren. Secure, privacy-preserving and federated machine learning in medical imaging. *Nature Machine Intelligence*, 2(6):305–311, 2020.

[127] T. Kauppi, V. Kalesnykiene, J.-K. Kamarainen, L. Lensu, I. Sorri, A. Raninen, R. Voutilainen, H. Uusitalo, H. Kälviäinen, and J. Pietilä. The diaretdb1 diabetic retinopathy database and evaluation protocol. In *BMVC*, volume 1, page 10. Citeseer, 2007.

[128] P. Kaur and P. Mahajan. Detection of brain tumors using a transfer learning-based optimized resnet152 model in mr images. *Computers in Biology and Medicine*, 178:109790, 2025.

[129] S. Key, M. Baygin, S. Demir, S. Dogan, and T. Tuncer. Meniscal tear and acl injury detection model based on alexnet and iterative relieff. *Journal of Digital Imaging*, 35(2):200–212, 2022.

[130] A. Khosravanian, M. Rahmanimanesh, P. Keshavarzi, S. Mozaffari, and K. Kazemi. Level set method for automated 3d brain tumor segmentation using symmetry analysis and kernel induced fuzzy clustering. *Multimedia Tools and Applications*, 1–22, 2022.

[131] P. Kickingereder, F. Isensee, I. Tursunova, J. Petersen, U. Neuberger, et al. Automated quantitative tumour response assessment of mri in neuro-oncology with artificial neural networks: a multicentre, retrospective study. *Lancet Oncology*, 20(5):728–740, 2019.

[132] D. H. Kim and T. Mackinnon. Artificial intelligence in fracture detection: transfer learning from deep convolutional neural networks. *Clinical Radiology*, 73(5):439–445, 2018.

[133] H. E. Kim, A. Cosa-Linan, N. Santhanam, M. Jannesari, M. E. Maros, and T. Ganslandt. Transfer learning for medical image classification: a literature review. *BMC Medical Imaging*, 22(1):69, 2022.

[134] S. Kollem. An efficient method for mri brain tumor tissue segmentation and classification using an optimized support vector machine. *Multimedia Tools and Applications*, 83(26):68487–68519, 2024.

[135] S. Korolev, A. Safiullin, M. Belyaev, and Y. Dodonova. Residual and plain convolutional neural networks for 3d brain mri classification. In *IEEE 14th International Symposium on Biomedical Imaging (ISBI)*, pages 835–838, Melbourne, VIC, Australia, 2017.

[136] M. Krichen. Generative adversarial networks. In *2023 14th International Conference on Computing Communication and Networking Technologies (ICCCNT)*, pages 1–7. IEEE, 2023.

[137] A. Krizhevsky, I. Sutskever, and G. E. Hinton. Imagenet classification with deep convolutional neural networks. In *Advances in Neural Information Processing Systems (NeurIPS)*, volume 25, pages 1097–1105, 2012.

[138] H. Kuang, Y. Wang, X. Tan, J. Yang, J. Sun, J. Liu, and Y. Chen. LW-CTrans: A lightweight hybrid network of CNN and Transformer for 3d medical image segmentation. *Medical Image Analysis*, 102:103545, 2025.

[139] K. Kumar, I. Suwalka, A. Uche-Ezennia, C. Iwendi, and C. N. Biamba. An improved deep learning unsupervised approach for mri tissue segmentation for alzheimer's disease detection. *IEEE Access*, 2024.

[140] K. A. Kumar and R. Boda. A computer-aided brain tumor diagnosis by adaptive fuzzy active contour fusion model and deep fuzzy classifier. *Multimedia Tools and Applications*, 1–37, 2022.

[141] R. Kumar, A. Gupta, H. S. Arora, and B. Raman. Ibrdm: An intelligent framework for brain tumor classification using radiomics-and dwt-based fusion of mri sequences. *ACM Transactions on Internet Technology (TOIT)*, 22(1):1–30, 2021.

[142] T. S. Kumar, C. Arun, and P. Ezhumalai. An approach for brain tumor detection using optimal feature selection and optimized deep belief network. *Biomedical Signal Processing and Control*, 73:103440, 2022.

[143] N. B. Kumarakulasinghe, T. Blomberg, J. Liu, A. S. Leao, and P. Papapetrou. Evaluating local interpretable model-agnostic explanations on clinical machine learning classification models. In *2020 IEEE 33rd International Symposium on Computer-Based Medical Systems (CBMS)*, September 2020.

[144] S. Kundu, S. Dutta, J. Mukhopadhyay, and N. Chakravorty. FFLUNet: Feature fused lightweight UNet for brain tumor segmentation. *Computers in Biology and Medicine*, 194:110460, 2025.

[145] Y. Lai, A. Cao, Y. Gao, J. Shang, Z. Li, and J. Guo. Advancing efficient brain tumor multi-class classification: New insights from the vision mamba model in transfer learning. *arXiv preprint arXiv: 2410.21872*, 2024.

[146] J. Lee, D. You, and H. Kim. Photometric transformer networks and label adjustment for breast density prediction. In *Proceedings of the IEEE/CVF International Conference on Computer Vision Workshop (ICCVW)*, pages 460–466, 2019.

[147] B. Lei, Z. Xia, F. Jiang, X. Jiang, Z. Ge, Y. Xu, et al. Skin lesion segmentation via generative adversarial networks with dual discriminators. *Medical Image Analysis*, 64:101716, 2020.

[148] G. Lepetit-Aimon, C. Playout, M. Carole Boucher, et al. Maples-dr: Messidor anatomical and pathological labels for explainable screening of diabetic retinopathy. *Scientific Data*, 2024.

[149] F. Li, X. Sheng, H. Wei, S. Tang, and H. Zou. Multi-lesion segmentation guided deep attention network for automated detection of diabetic retinopathy. *Computers in Biology and Medicine*, 183:109352, 2024.

[150] Z. Li, J. Zhang, T. Tan, X. Teng, X. Sun, H. Zhao, L. Liu, Y. Xiao, B. Lee, Y. Li, et al. Deep learning methods for lung cancer segmentation in whole-slide histopathology images—the acdc@ lunghp challenge 2019. *IEEE Journal of Biomedical and Health Informatics*, 25(2):429–440, 2020.

[151] A. Liew, C. C. Lee, B. L. Lan, and M. Tan. CASPIANET++: A multidimensional channel-spatial asymmetric attention network with noisy student curriculum learning paradigm for brain tumor segmentation. *Computers in Biology and Medicine*, 136:104690, 2021.

[152] C. W. Lin and Z. Chen. MM-UNet: A novel cross-attention mechanism between modules and scales for brain tumor segmentation. *Engineering Applications of Artificial Intelligence*, 133:108591, 2024.

[153] T. Lin, P. Dollár, R. Girshick, K. He, B. Hariharan, and S. Belongie. Feature pyramid networks for object detection. In *Proceedings of the IEEE Conference on Computer Vision and Pattern Recognition*, pages 2117–2125, 2017.

[154] T.-Y. Lin, P. Goyal, R. Girshick, K. He, and P. Dollár. Focal loss for dense object detection. In *Proceedings of the IEEE International Conference on Computer Vision*, pages 2980–2988, 2017.

[155] F. Liu, B. Guan, Z. Zhou, A. Samsonov, H. Rosas, K. Lian, R. Sharma, A. Kanarek, J. Kim, A. Guermazi, et al. Fully automated diagnosis of anterior cruciate ligament tears on knee mr images by using deep learning. *Radiology. Artificial Intelligence*, 1(3):180091, 2019.

[156] J. Liu, M. Li, Y. Luo, S. Yang, W. Li, and Y. Bi. Alzheimer's disease detection using depthwise separable convolutional neural networks. *Computer Methods and Programs in Biomedicine*, 203:106032, 2021.

[157] J. Liu, J. Zheng, and G. Jiao. Transition net: 2d backbone to segment 3d brain tumor. *Biomedical Signal Processing and Control*, 75:103622, 2022.

[158] S. Liu, A. V. Masurkar, H. Rusinek, J. Chen, B. Zhang, W. Zhu, C. Fernandez-Granda, and N. Razavian. Generalizable deep learning model for early alzheimer's disease detection from structural mris. *Scientific Reports*, 12(1):17106, 2022.

[159] Y. Liu, J. Du, C. M. Vong, G. Yue, J. Yu, Y. Wang, and T. Wang. Scale-adaptive super-feature based MetricUNet for brain tumor segmentation. *Biomedical Signal Processing and Control*, 73:103442, 2022.

[160] Y. Liu, Y. Ma, Z. Zhu, J. Cheng, and X. Chen. Transsea: Hybrid cnn–transformer with semantic awareness for 3-d brain tumor segmentation. *IEEE Transactions on Instrumentation and Measurement*, 73:16–31, 2024.

[161] Z. Liu, Y. Lin, Y. Cao, H. Hu, Y. Wei, Z. Zhang, S. Lin, and B. Guo. Swin transformer: Hierarchical vision transformer using shifted windows. In *Proceedings of the IEEE/CVF International Conference on Computer Vision*, pages 9992–10002, 2021.

[162] N.-H. Lu, Y.-H. Huang, K.-Y. Liu, and T.-B. Chen. Deep learning-driven brain tumor classification and segmentation using non-contrast MRI. *Scientific Reports*, 15(27831):1–17, 2025.

[163] S. Lundberg and S.-I. Lee. A unified approach to interpreting model predictions. https://doi.org/10.48550/arXiv.1705.07874, 2017. arXiv:1705.07874 [cs.AI].

[164] M. Lundblad, M. Englund, M.-R. Järvelin, and S. Lohmander. Anterior cruciate ligament injury and radiologic progression of knee osteoarthritis: a systematic review and meta-analysis. *American Journal of Sports Medicine*, 41(10):2242–2254, 2013.

[165] E. A. Lundeen, Z. Burke-Conte, D. B. Rein, J. S. Wittenborn, J. Saaddine, A. Y. Lee, and A. D. Flaxman. Prevalence of diabetic retinopathy in the us in 2021. *JAMA Ophthalmology*, 141(8):747–754, 2023.

[166] L. Maier-Hein, M. Eisenmann, A. Reinke, S. Burghart, M. Giannoni, Tobias, et al. Roß. Isles 2015—a public evaluation benchmark for stroke lesion segmentation from multimodal mri. *Medical Image Analysis*, 35:250–269, 2017.

[167] D. Maji, P. Sigedar, and M. Singh. Attention res-unet with guided decoder for semantic segmentation of brain tumors. *Biomedical Signal Processing and Control*, 71:103077, 2022.

[168] P. K. Mallick, S. H. Ryu, S. K. Satapathy, S. Mishra, G. N. Nguyen, and P. Tiwari. Brain mri image classification for cancer detection using deep wavelet autoencoder-based deep neural network. *IEEE Access*, 7:46278–46287, 2019.

[169] S. K. Mamatha, H. K. Krishnappa, and N. Shalini. Graph theory based segmentation of magnetic resonance images for brain tumor detection. *Pattern Recognition and Image Analysis*, 32(1):153–161, 2022.

[170] U. Mandawkar and T. Diwan. Hybrid cuttle fish-grey wolf optimization tuned weighted ensemble classifier for alzheimer's disease classification. *Biomedical Signal Processing and Control*, 92:106101, 2024.

[171] K. L. Markolf, D. M. Burchfield, M. S. Shapiro, M. F. Shepard, G. A. M. Finerman, and J. L. Slauterbeck. Effects of femoral tunnel placement on knee laxity and forces in an anterior cruciate ligament graft. *The Journal of Bone and Joint Surgery*, 77(7):1058–1066, 1995.

[172] H. B. McMahan, E. Moore, D. Ramage, S. Hampson, and B. A. y Arcas. Communication-efficient learning of deep networks from decentralized data. In *AISTATS*, 2017.

[173] T. Meena and S. Roy. Bone fracture detection using deep supervised learning from radiological images: A paradigm shift. *Diagnostics*, 12(10):2420, 2022.

[174] X. Mei, Z. Liu, P. M. Robson, B. Marinelli, M. Huang, A. Doshi, A. Jacobi, C. Cao, K. E. Link, T. Yang, Y. Wang, H. Greenspan, T. Deyer, Z. A. Fayad, and Y. Yang. Radimagenet: An open radiologic deep learning research dataset for effective transfer learning. *Radiology. Artificial Intelligence*, 4(5):e210315, 2022.

[175] T. Mendonça, P. M. Ferreira, J. S. Marques, A. R. S. Marcal, and J. Rozeira. Ph 2-a dermoscopic image database for research and benchmarking. In *2013 35th annual international conference of the IEEE engineering in medicine and biology society (EMBC)*, pages 5437–5440. IEEE, 2013.

[176] T. Mendonça, P. M. Ferreira, J. S. Marques, A. R. S. Marçal, and J. Rozeira. Ph2 - a public database for the analysis of dermoscopic images. In *EMBC*, pages 5437–5440, 2013.

[177] B. H. Menze, A. Jakab, S. Bauer, J. Kalpathy-Cramer, K. Farahani, J. Kirby, Y. Burren, N. Porz, J. Slotboom, R. Wiest, et al. The multimodal brain tumor image segmentation benchmark (brats). *IEEE Transactions on Medical Imaging*, 34(10):1993–2024, 2014.

[178] K. D. Miller, L. Nogueira, A. B. Mariotto, J. H. Rowland, K. R. Yabroff, C. M. Alfano, A. Jemal, J. L. Kramer, and R. L. Siegel. Cancer treatment and survivorship statistics, 2019. *CA: A Cancer Journal for Clinicians*, 69(5):363–385, 2019.
[179] H. Min, Y. Rabi, A. Wadhawan, P. Bourgeat, J. Dowling, J. White, A. Tchernegovski, B. Formanek, M. Schuetz, G. Mitchell, et al. Automatic classification of distal radius fracture using a two-stage ensemble deep learning framework. *Physical and Engineering Sciences in Medicine*, 46(2):877–886, 2023.
[180] R. Mondal, N. Ray, K. Sheets, R. Liao, V. Singh, and S. Jha. Deep learning for cardiac image segmentation: A review. *Frontiers in Cardiovascular Medicine*, 5:164, 2018.
[181] J. Morano, Á. S. Hervella, J. Novo, and J. Rouco. Simultaneous segmentation and classification of the retinal arteries and veins from color fundus images. *Artificial Intelligence in Medicine*, 118:102116, 2021.
[182] I. C. Moreira, I. Amaral, I. Domingues, A. Cardoso, M. J. Cardoso, and J. S. Cardoso. Inbreast: toward a full-field digital mammographic database. *Academic Radiology*, 19(2):236–248, 2012.
[183] M. A. Morid, A. Borjali, and F. Faghri. Multistage transfer learning for medical images. *Artificial Intelligence Review*, 57(8):1–25, 2024.
[184] M. F. Mridha, M. A. Hamid, M. M. Monowar, A. J. Keya, A. Q. Ohi, M. R. Islam, and J.-M. Kim. A comprehensive survey on deep-learning-based breast cancer diagnosis. *Cancers*, 13:6116, 2021.
[185] T. Murakami, K. Takahashi, T. Ishihara, T. Kawashima, T. Okazaki, S. Ikemura, T. Kumagai, and H. Fujita. Anterior cruciate ligament tear detection using convolutional neural network trained on the mrnet dataset. *BMC Musculoskeletal Disorders*, 23:120, 2023.
[186] K. Murphy, J. Hoffman, L. Prevedello, Y. A. Song, P. Siva, et al. Artificial intelligence for medical image analysis: A guide for authors and reviewers. *American Journal of Roentgenology*, 211(3):505–515, 2018.
[187] A. Myronenko. 3d mri brain tumor segmentation using autoencoder regularization. In *International MICCAI Brainlesion Workshop*, pages 311–320. Springer, 2018.
[188] A. Myronenko and A. Hatamizadeh. Robust semantic segmentation of brain tumor regions from 3d mris. In *International MICCAI Brainlesion Workshop*, pages 82–89. Springer, 2019.
[189] G. Mårtensson, D. Ferreira, T. Granberg, and E. Westman. The reliability of a deep learning model in clinical out-of-distribution mri data: a multicohort study. *arXiv preprint arXiv:1911.00515*, 2019.
[190] J. Nalepa, S. Adamski, K. Kotowski, S. Chelstowska, M. Machnikowska-Sokolowska, O. Bozek, A. Wisz, and E. Jurkiewicz. Segmenting pediatric optic pathway gliomas from mri using deep learning. *Computers in Biology and Medicine*, 142:105237, 2022.
[191] N. K. Namiri, B. Astuto, T. M. Link, V. Pedoia, and S. Majumdar. Hierarchical severity staging of anterior cruciate ligament injuries using deep learning with mri images. *Radiology. Artificial Intelligence*, 2(5):e190207, 2020.
[192] M. Napravnik, F. Hržić, M. Urschler, D. Miletić, and I. Štajduhar. Lessons learned from radiologynet foundation models for transfer learning in medical radiology. *Scientific Reports*, 15(1):21622, 2025.
[193] A. Narin, C. Kaya, and Z. Pamuk. Automatic detection of coronavirus disease (covid-19) using x-ray images and deep convolutional neural networks. *Pattern Recognition Letters*, 131:116–121, 2020.
[194] G. Neelima, D. R. Chigurukota, B. Maram, and B. Girirajan. Optimal deepmrseg based tumor segmentation with gan for brain tumor classification. *Biomedical Signal Processing and Control*, 74:103537, 2022.
[195] D. Nie, R. Trullo, J. Lian, C. Petitjean, S. Ruan, Q. Wang, and D. Shen. Medical image synthesis with context-aware generative adversarial networks. *Medical Image Computing and Computer-Assisted Intervention (MICCAI)*, pages 417–425, 2017.
[196] F. R. Noyes and S. D. Barber. Arthroscopy in the diagnosis and treatment of knee disorders. *Arthroscopy: The Journal of Arthroscopic & Related Surgery*, 7(1):78–95, 1991.
[197] O. Oktay, J. Schlemper, L. L. Folgoc, M. Lee, M. Heinrich, K. Misawa, et al. Attention u-net: Learning where to look for the pancreas. *arXiv preprint*, 2018.
[198] O. Oktay, J. Schlemper, L. Le Folgoc, M. Lee, M. Heinrich, K. Misawa, K. Mori, S. McDonagh, N. Y. Hammerla, B. Kainz, et al. Attention u-net: Learning where to look for the pancreas. *arXiv preprint arXiv:1804.03999*, 2018.

[199] J. I. Orlando, E. Prokofyeva, M. Del Fresno, and M. B. Blaschko. An ensemble deep learning based approach for red lesion detection in fundus images. *Computer Methods and Programs in Biomedicine*, 153:115–127, 2018.

[200] N. Orlando, D. J. Gillies, I. Gyacskov, C. Romagnoli, D. D'Souza, and A. Fenster. Automatic prostate segmentation using deep learning on clinically diverse 3d transrectal ultrasound images. *Medical Physics*, 47(6):2413–2426, 2020.

[201] X. Ouyang and et al.. Contrastive self-supervised learning for diabetic retinopathy early detection. *Scientific Reports*, 13:5559, 2023.

[202] I. Pacal. A novel swin transformer approach utilizing residual multi-layer perceptron for diagnosing brain tumors in mri images. *International Journal of Machine Learning and Cybernetics*, 15:3579–3597, 2024.

[203] S. Pachade, P. Porwal, M. Kokare, L. Giancardo, and F. Mériaudeau. Nenet: Nested efficientnet and adversarial learning for joint optic disc and cup segmentation. *Medical Image Analysis*, 74:102253, 2021.

[204] D. Pan, J. Shen, Z. Al-Huda, and M. A. A. Al-Qaness. Vcanet: Vision transformer with fusion channel and spatial attention module for 3d brain tumor segmentation. *Computers in Biology and Medicine*, 186:109662, 2025.

[205] S. J. Pan and Q. Yang. A survey on transfer learning. *IEEE Transactions on Knowledge and Data Engineering*, 22(10):1345–1359, 2010.

[206] V. C. Pangman, J. Sloan, and L. Guse. An examination of psychometric properties of the mini-mental state examination and the standardized mini-mental state examination: Implications for clinical practice. *Applied Nursing Research*, 13(4):209–213, 2000.

[207] K. Pani and I. Chawla. A hybrid approach for multi modal brain tumor segmentation using two phase transfer learning, SSL and a hybrid 3DUNET. *Computers & Electrical Engineering*, 118:109418, 2024.

[208] M. S. Park and D. H. Lee. Challenges in automated acl injury detection from mri: A review. *European Radiology*, 31(1):236–245, 2021.

[209] C. Patrício, J. C. Neves, and L. F. Teixeira. Survey of explainable artificial intelligence techniques for biomedical imaging with deep neural networks. *Computers in Biology and Medicine*, 156:106668, 2023.

[210] H. Peiris, M. Hayat, Z. Chen, G. Egan, and M. Harandi. Hybrid window attention based transformer architecture for brain tumor segmentation. *arXiv preprint arXiv:2209.07704*, 2022.

[211] S. Pereira, A. Pinto, V. Alves, and C. A. Silva. Brain tumor segmentation using convolutional neural networks in mri images. *IEEE Transactions on Medical Imaging*, 35(5):1240–1251, 2016.

[212] P. Porwal, S. Pachade, R. Kamble, M. Kokare, G. Deshmukh, V. Sahasrabuddhe, and F. Meriaudeau. Indian diabetic retinopathy image dataset (idrid): a database for diabetic retinopathy screening research. *Data*, 3(3):25, 2018.

[213] A. Pourmahboubi, N. Arsalani Saeed, and H. Tabrizchi. A brain tumor segmentation enhancement in mri images using u-net and transfer learning. *BMC Medical Imaging*, 25(307):1–24, 2025.

[214] B. S. Prabakaran, P. Hamelmann, E. Ostrowski, and M. Shafique. Fpus23: an ultrasound fetus phantom dataset with deep neural network evaluations for fetus orientations, fetal planes, and anatomical features. *IEEE Access*, 2023.

[215] H. Pratt, F. Coenen, D. M. Broadbent, S. P. Harding, and Y. Zheng. Convolutional neural networks for diabetic retinopathy. *Procedia Computer Science*, 90:200–205, 2016.

[216] J. Pruthi, S. Arora, and K. Khanna. Brain tumor segmentation using river formation dynamics and active contour model in magnetic resonance images. *Neural Computing & Applications*, 1–10, 2022.

[217] B. Pu, K. Li, S. Li, and N. Zhu. Automatic fetal ultrasound standard plane recognition based on deep learning and iiot. *IEEE Transactions on Industrial Informatics*, 17(11):7771–7780, 2021.

[218] A. Puente-Castro, E. Fernandez-Blanco, A. Pazos, and C. R. Munteanu. Automatic assessment of alzheimer's disease diagnosis based on deep learning techniques. *Computers in Biology and Medicine*, 120:103764, 2020.

[219] C. Qin, W. Li, B. Zheng, J. Zeng, S. Liang, X. Zhang, and W. Zhang. Dual adversarial models with cross-coordination consistency constraint for domain adaption in brain tumor segmentation. *Frontiers in Neuroscience*, 17:1043533, 2023.

[220] G. Quellec, K. Charrière, Y. Boudi, B. Cochener, and M. Lamard. Deep image mining for diabetic retinopathy screening. *Medical Image Analysis*, 39:178–193, 2017.

[221] I. Qureshi, J. Ma, and Q. Abbas. Diabetic retinopathy detection and stage classification in eye fundus images using active deep learning. *Multimedia Tools and Applications*, 80(8):11691–11721, 2021.

[222] M. Raghu, C. Zhang, J. Kleinberg, and S. Bengio. Transfusion: Understanding transfer learning for medical imaging. In *Advances in Neural Information Processing Systems (NeurIPS)*, pages 3347–3357, 2019.

[223] T. Rahman, A. Khandakar, Y. Qiblawey, A. Tahir, S. Kiranyaz, S. B. A. Kashem, M. T. Islam, S. Al Maadeed, S. M. Zughaier, M. S. Khan, et al. Exploring the effect of image enhancement techniques on covid-19 detection using chest x-ray images. *Computers in Biology and Medicine*, 132:104319, 2021.

[224] H. M. Rai, K. Chatterjee, and S. Dashkevich. Automatic and accurate abnormality detection from brain mr images using a novel hybrid UnetResNext-50 deep CNN model. *Biomedical Signal Processing and Control*, 66:102477, 2021.

[225] P. Rajpurkar, J. Irvin, K. Zhu, B. Yang, H. Mehta, T. Duan, D. Ding, A. Bagul, C. Langlotz, K. Shpanskaya, et al. Chexnet: Radiologist-level pneumonia detection on chest x-rays with deep learning. *arXiv preprint arXiv:1711.05225*, 2017.

[226] C. S. Rao and K. Karunakara. Efficient detection and classification of brain tumor using kernel based svm for mri. *Multimedia Tools and Applications*, 1–25, 2022.

[227] J. Raya, N. Bien, P. Pelissier, P. Guyon, E. Ackermann, S. Laporte, L. Bourrelier, V. Nguyen, I. Buvat, D. Hacquin, H. Noirot, M. Laborde, P. Goupille, J. Azais, M. Cohen, P. García March, F. Campillo-Navarro, J. Dolz, M. Piérard, A. Vachon, C. Barbare, F. Montagne, P. Cardot, M. Akkari, J.-Y. Navez, S. Laporte, F. Barelli, F. Pilleul, B. Poncet, S. Ryu, P. Noël, A. Vaitunin, A. Tiran, A. Bernard, and N. Ayache. Deep learning to detect anterior cruciate ligament tear on knee mri: multi-continental external validation. *Radiology. Artificial Intelligence*, 4(3):e220043, 2022.

[228] C. K. K. Reddy, P. A. Reddy, H. Janapati, B. Assiri, M. Shuaib, S. Alam, and A. Sheneamer. A fine-tuned vision transformer based enhanced multi-class brain tumor classification using MRI scan imagery. *Frontiers in Oncology*, 14:1400341, 2024.

[229] REFUGE2-2020. Retinal fundus glaucoma challenge edition 2, 2020.

[230] T. Ren, E. Honey, H. Rebala, A. Sharma, A. Chopra, and M. Kurt. An optimization framework for processing and transfer learning for the brain tumor segmentation. *arXiv preprint arXiv:2402.07008*, 2024.

[231] A. Rodtook, K. Kirimasthong, W. Lohitvisate, and S. S. Makhanov. Automatic initialization of active contours and level set method in ultrasound images of breast abnormalities. *Pattern Recognition*, 79:172–182, 2018.

[232] O. Ronneberger, P. Fischer, and T. Brox. U-net: Convolutional networks for biomedical image segmentation. In *International Conference on Medical Image Computing and Computer-Assisted Intervention*, page 234–241. Springer, 2015.

[233] V. Rotemberg, N. Kurtansky, B. Betz-Stablein, L. Caffery, E. Chousakos, N. Codella, M. Combalia, S. Dusza, P. Guitera, D. Gutman, et al. A patient-centric dataset of images and metadata for identifying melanomas using clinical context. *Scientific Data*, 8(1):34, 2021.

[234] T. C. Roth, H. Zollinger, A. B. Imhoff, and W. Kucharczyk. Mr imaging of the anterior cruciate ligament and associated injuries. *Acta Radiologica*, 43(5):511–519, 2002.

[235] S. S. Roy, R. Sikaria, and A. Susan. A deep learning based cnn approach on mri for alzheimer's disease detection. *Intelligent Decision Technologies*, 13(4):495–505, 2019.

[236] P. Ruamviboonsuk, J. Krause, P. Chotcomwongse, R. Sayres, R. Raman, K. Widner, B. J. Campana, S. Phene, K. Hemarat, M. Tadarati, N. Kitnarong, K. Boyer, T. Rungsilp, A. Y. Lee, L. Peng, D. Webster, and V. Gulshan. Deep learning versus human graders for classifying diabetic retinopathy severity in a nationwide screening programme. *npj Digital Medicine*, 2:25, 2019.

[237] P. Ruamviboonsuk, R. Tiwari, R. Sayres, V. Nganthavee, K. Hemarat, A. Kongprayoon, R. Raman, B. Levinstein, Y. Liu, M. Schaekermann, R. Lee, S. Virmani, K. Widner, J. Chambers, F. Hersch, L. Peng, and R. D. Webster. Real-time diabetic retinopathy screening by deep learning in a multisite national screening programme: a prospective interventional cohort study. *The Lancet Digital Health*, 4(4):e235–e244, 2022.

[238] D. A. Rubin, J. M. Kettering, J. D. Towers, and C. A. Britton. Mr imaging of the knee: sports-related injuries. *Radiology*, 219(3):835–842, 2001.

[239] F. Rustom, E. Moroze, P. Parva, H. Ogmen, and A. Yazdanbakhsh. Deep learning and transfer learning for brain tumor detection and classification. *Biology Methods and Protocols*, 9(1):bpae080, 2024.

[240] N. Sambyal, P. Saini, R. Syal, and V. Gupta. Modified u-net architecture for semantic segmentation of diabetic retinopathy images. *Biocybernetics and Biomedical Engineering*, 40(3):1094–1109, 2020.

[241] M. M. K. Sarker, H. A. Rashwan, F. Akram, V. K. Singh, S. F. Banu, F. U. H. Chowdhury, K. A. Choudhury, S. Chambon, P. Radeva, D. Puig, et al. Slsnet: Skin lesion segmentation using a lightweight generative adversarial network. *Expert Systems with Applications*, 183:115433, 2021.

[242] V. V. S. Sasank and S. Venkateswarlu. An automatic tumour growth prediction based segmentation using full resolution convolutional network for brain tumour. *Biomedical Signal Processing and Control*, 71:103090, 2022.

[243] L. Scarpace, T. Mikkelsen, S. Cha, et al. Radiology data from the cancer genome atlas glioblastoma multiforme [tcga-gbm] collection. *Cancer Imaging Archive*, 11(4):1, 2016.

[244] J. Schlemper, O. Oktay, L. Chen, J. Matthew, C. Knight, B. Kainz, et al. Attention-gated networks for improving ultrasound scan plane detection. *arXiv preprint*, 2018.

[245] D. Schub and P. Saluan. Current concepts in the evaluation and treatment of the anterior cruciate ligament. *The Orthopedic Clinics of North America*, 40(1):47–66, 2009.

[246] R. R. Selvaraju, M. Cogswell, A. Das, R. Vedantam, D. Parikh, and D. Batra. Grad-cam: Visual explanations from deep networks via gradient-based localization. In *Proceedings of the IEEE International Conference on Computer Vision (ICCV)*, pages 618–626, 2017.

[247] A. A. A. Setio, A. Traverso, T. de Bel, M. S. Berens, C. van den Bogaard, P. Cerello, H. Chen, Q. Dou, M. E. Fantacci, H. Geurts, et al. Validation, comparison, and combination of algorithms for automatic detection of pulmonary nodules in computed tomography images: the luna16 challenge. *Medical Image Analysis*, 42:1–13, 2017.

[248] B. R. Setty, K. Vishwanath, G. J. Puneeth, and D. B. Sreepathi. Survey on features and techniques used for bone fracture detection and classification. *International Research Journal of Engineering and Technology*, 7(5):1585–1594, 2020.

[249] K. Shaheed, A. Mao, I. Qureshi, M. Kumar, S. Hussain, and X. Zhang. Recent advancements in finger vein recognition technology: methodology, challenges and opportunities. *Information Fusion*, 79:84–109, 2022.

[250] Z. Shahvaran, K. Kazemi, M. Fouladivanda, M. S. Helfroush, O. Godefroy, and A. Aarabi. Morphological active contour model for automatic brain tumor extraction from multimodal magnetic resonance images. *Journal of Neuroscience Methods*, 362:109296, 2021.

[251] F. E. Shamout, S. Cox, B. W. Lee, T. Dykstra, G. Shih, B. Reig, and K. J. Geras. Artificial intelligence system reduces false-positive findings in the interpretation of breast ultrasound exams. *Nature Communications*, 11(1):5645, 2020.

[252] M. Sharif, J. Amin, M. Raza, M. A. Anjum, H. Afzal, and S. A. Shad. Brain tumor detection based on extreme learning. *Neural Computing & Applications*, 32(20):15975–15987, 2020.

[253] U. Sharif and Z. Aslam. Evaluation of semi-automated and automated methods in acl injury detection. *Computers in Biology and Medicine*, 138:104989, 2021.

[254] A. K. Sharma, A. Nandal, A. Dhaka, L. Zhou, A. Alhudhaif, F. Alenezi, and K. Polat. Brain tumor classification using the modified resnet50 model based on transfer learning. *Biomedical Signal Processing and Control*, 86:105299, 2023.

[255] N. Sharma and P. Lalwani. A multi model deep net with an explainable ai based framework for diabetic retinopathy segmentation and classification. *Scientific Reports*, 15(1):8777, 2025.
[256] M. J. Sheller, B. Edwards, G. A. Reina, J. Martin, S. Pati, J. R. Kotrotsou, D. B. Marcus, T. D. N. George, and S. Bakas. Federated learning in medicine: Facilitating multi-institutional collaborations without sharing patient data. *Scientific Reports*, 10(1):12598, 2020.
[257] W. Shen and F. Zhang. Automated acl tear diagnosis using cnns on mri. *IEEE Transactions on Medical Imaging*, 38(8):2007–2016, 2019.
[258] Z. Shi, D. Wang, C. Han, C. Liang, and Z. Liu. Dataset for lung tumor segmentation on ct images through federated semi-supervised with dynamic update strategy. CT-Scans, 2024.
[259] H.-C. Shin, H. R. Roth, M. Gao, L. Lu, Z. Xu, I. Nogues, J. Yao, D. Mollura, and R. M. Summers. Deep convolutional neural networks for computer-aided detection: CNN architectures, dataset characteristics and transfer learning. *IEEE Transactions on Medical Imaging*, 35(5):1285–1298, 2016.
[260] N. V. Shree and T. N. R. Kumar. Identification and classification of brain tumor mri images with feature extraction using dwt and probabilistic neural network. *Brain Informatics*, 5(1):23–30, 2018.
[261] S. Shurrab and R. Duwairi. Self-supervised learning methods and applications in medical imaging: A survey. *PeerJ Computer Science*, 7:e564, 2021.
[262] A. A. Siddique, A. Raza, M. S. Alshehri, N. Alasbali, and S. F. Abbasi. Optimizing tumor classification through transfer learning and particle swarm optimization-driven feature extraction. *IEEE Access*, 12:85929–85939, 2024.
[263] G. Silva, L. Oliveira, and M. Pithon. Automatic segmenting teeth in x-ray images: Trends, a novel data set, benchmarking and future perspectives. *Expert Systems with Applications*, 107:15–31, 2018.
[264] K. Simonyan and A. Zisserman. Very deep convolutional networks for large-scale image recognition. In *Proceedings of the International Conference on Learning Representations*, 2015.
[265] J. Sivaswamy, S. R. Krishnadas, G. D. Joshi, M. Jain, and A. U. S. Tabish. Drishti-gs: Retinal image dataset for optic nerve head (onh) segmentation. In *2014 IEEE 11th international symposium on biomedical imaging (ISBI)*, pages 53–56. IEEE, 2014.
[266] A. Skouta, A. Elmoufidi, S. Jai-Andaloussi, and O. Ouchetto. Hemorrhage semantic segmentation in fundus images for the diagnosis of diabetic retinopathy by using a convolutional neural network. *Journal of Big Data*, 9(1):78, 2022.
[267] M. Soltaninejad, G. Yang, T. Lambrou, N. Allinson, T. L. Jones, T. R. Barrick, F. A. Howe, and X. Ye. Supervised learning based multimodal mri brain tumour segmentation using texture features from supervoxels. *Computer Methods and Programs in Biomedicine*, 157:69–84, 2018.
[268] R. F. Spaide, J. G. Fujimoto, N. K. Waheed, S. R. Sadda, and G. Staurenghi. Optical coherence tomography angiography. *Progress in Retinal and Eye Research*, 64:1–55, 2018.
[269] C. Spampinato, S. Palazzo, D. Giordano, M. Aldinucci, and R. Leonardi. Deep learning for automated skeletal bone age assessment in x-ray images. *Medical Image Analysis*, 36:41–51, 2017.
[270] F. A. Spanhol, L. S. Oliveira, C. Petitjean, and L. Heutte. Breast cancer histopathological image classification using convolutional neural networks. In *Neural Networks (IJCNN), 2016 International Joint Conference on*, pages 2560–2567. IEEE, 2016.
[271] R. A. Sperling, M. C. Donohue, R. Raman, et al. Trial of solanezumab in preclinical alzheimer's disease. *The New England Journal of Medicine*, 389(12):1096–1107, 2023.
[272] S. Sridevi and A. RajivKannan. Development of 3dtdunet++ with novel function and multi-scale dilated-based deep learning model for lung cancer diagnosis using ct images. *Biomedical Signal Processing and Control*, 94:106243, 2024.
[273] J. Staal, M. D. Abràmoff, M. Niemeijer, M. A. Viergever, and B. Van Ginneken. Ridge-based vessel segmentation in color images of the retina. *IEEE Transactions on Medical Imaging*, 23(4):501–509, 2004.
[274] Y. Su, Q. Liu, W. Xie, and P. Hu. Yolo-logo: A transformer-based yolo segmentation model for breast mass detection and segmentation in digital mammograms. *Computer Methods and Programs in Biomedicine*, 221:106903, 2022.

[275] H.-I. Suk, S.-W. Lee, D. Shen, and The Alzheimer's Disease Neuroimaging Initiative. Deep sparse multi-task learning for feature selection in Alzheimer's disease diagnosis. *Brain Structure and Function*, 221(5):2569–2587, 2016.

[276] J. Sun, Y. Peng, Y. Guo, and D. Li. Segmentation of the multimodal brain tumor image used the multi-pathway architecture method based on 3d fcn. *Neurocomputing*, 423:34–45, 2021.

[277] L. Sun et al. Contrastive learning-based pretraining improves representation for rib fracture detection in chest radiographs. *Scientific Reports*, 13:10234, 2023.

[278] M. Sun, K. Li, X. Qi, H. Dang, and G. Zhang. Contextual information enhanced convolutional neural networks for retinal vessel segmentation in color fundus images. *Journal of Visual Communication and Image Representation*, 77:103134, 2021.

[279] M. Sun, X. Li, and W. Sun. Image generation and lesion segmentation of brain tumors and stroke based on gan and 3d resu-net. *IEEE Access*, 2024.

[280] Y. Sun and C. Wang. A computation-efficient CNN system for high-quality brain tumor segmentation. *Biomedical Signal Processing and Control*, 74:103475, 2022.

[281] H. Sung, J. Ferlay, R. L. Siegel, M. Laversanne, I. Soerjomataram, A. Jemal, and F. Bray. Global cancer statistics 2020: Globocan estimates of incidence and mortality worldwide for 36 cancers in 185 countries. *CA: A Cancer Journal for Clinicians*, 71(3):209–249, 2021.

[282] I. Suwalka and N. Agrawal. An improved unsupervised mapping technique using amsom for neurodegenerative disease detection. *International Journal of Computer Systems Engineering*, 4(2–3):185–194, 2018.

[283] Z. N. K. Swati, Q. Zhao, M. Kabir, F. Ali, Z. Ali, S. Ahmed, and J. Lu. Content-based brain tumor retrieval for mr images using transfer learning. *IEEE Access*, 7:17809–17822, 2019.

[284] C. Szegedy, W. Liu, Y. Jia, P. Sermanet, S. E. Reed, D. Anguelov, D. Erhan, V. Vanhoucke, and A. Rabinovich. Going deeper with convolutions. In *Proceedings of the IEEE Conference on Computer Vision and Pattern Recognition (CVPR)*, pages 1–9, 2015.

[285] A. Tahir, A. Saadia, K. Khan, A. Gul, A. Qahmash, and R. N. Akram. Enhancing diagnosis: ensemble deep-learning model for fracture detection using x-ray images. *Clinical Radiology*, 79(11):e1394–e1402, 2024. e1402.

[286] N. Tajbakhsh, J. Y. Shin, S. R. Gurudu, R. V. Hurst, J. B. Kendall, M. B. Gotway, and J. Liang. Convolutional neural networks for medical image analysis: Full training or fine tuning? *IEEE Transactions on Medical Imaging*, 35(5):1299–1312, 2016.

[287] L. F. Talib, J. Amin, M. Sharif, and M. Raza. Transformer-based semantic segmentation and cnn network for detection of histopathological lung cancer. *Biomedical Signal Processing and Control*, 92:106106, 2024.

[288] M. Tan and Q. V. Le. Efficientnet: Rethinking model scaling for convolutional neural networks. In *Proceedings of the 36th International Conference on Machine Learning (ICML)*, pages 6105–6114, 2019.

[289] P. Tang, X. Yan, Q. Liang, and D. Zhang. Afln-dgcl: Adaptive feature learning network with difficulty-guided curriculum learning for skin lesion segmentation. *Applied Soft Computing*, 110:107656, 2021.

[290] A. Tariq, M. M. Iqbal, M. J. Iqbal, and I. Ahmad. Transforming brain tumor detection: Empowering multi-class classification with vision transformers and efficientnetv2. *IEEE Access*, 13:63857–63873, 2025.

[291] S. Tayebi Arasteh and et al.. Enhancing diagnostic deep learning via self-supervised pretraining on large-scale, unlabeled non-medical images. *European Radiology Experimental*, 8, 2024.

[292] Z. L. Teo, Y.-C. Tham, M. Yu, M. L. Chee, T. H. Rim, N. Cheung, M. M. Bikbov, Y. X. Wang, Y. Tang, Y. Lu, I. Y. Wong, D. S. W. Ting, G. S. W. Tan, J. B. Jonas, C. Sabanayagam, T. Y. Wong, and C.-Y. Cheng. Global prevalence of diabetic retinopathy and projection of burden through 2045: Systematic review and meta-analysis. *Ophthalmology*, 128(11):1580–1591, 2021.

[293] Y. L. Thian, Y. Li, P. Jagmohan, D. Sia, V. E. Y. Chan, and R. T. Tan. Fracture detection and localization on wrist radiographs. *Radiology. Artificial Intelligence*, 1(1):e180001, 2019.

[294] D. S. W. Ting, C. Y. Cheung, G. Lim, G. S. W. Tan, N. Quang, A. Gan, H. Hamzah, R. Garcia-Franco, I. Y. S. Yeo, S. Y. Lee, and T. Y. Wong. Development and validation of a deep learning system for diabetic retinopathy and related eye diseases using retinal images from multi-ethnic populations with diabetes. *JAMA*, 318(22):2211–2223, 2017.

[295] A. Tiulpin, J. Thevenot, E. Rahtu, P. Lehenkari, and S. Saarakkala. Automatic knee osteoarthritis diagnosis from plain radiographs: a deep learning-based approach. *Scientific Reports*, 8(1):1727, 2018.

[296] M. A. Tokgoz, E. B. Oklaz, O. Ak, E. B. Guler Oklaz, and M. B. Ataoglu. The potential of posterior cruciate ligament buckling phenomenon as a sign for partial anterior cruciate ligament tears. *Archives of Orthopaedic and Trauma Surgery*, 144:2181–2187, 2024.

[297] T. Tong, K. Gray, Q. Gao, L. Chen, and D. Rueckert. Multi-modal classification of Alzheimer's disease using nonlinear graph fusion. *Pattern Recognition*, 63(1):171–181, 2017.

[298] J. S. Torg, W. Conrad, and V. Kalen. Assessment of the function of the anterior cruciate ligament in the athlete. *American Journal of Sports Medicine*, 19(3):238–243, 1991.

[299] J. A. Trejo-Lopez, A. T. Yachnis, and S. Prokop. Neuropathology of alzheimer's disease. *Neurotherapeutics*, 19(1):173–185, 2023.

[300] T. Truong, S. Mohammadi, M. Lenga, et al. How transferable are self-supervised features in medical image classification tasks? *Proceedings of Machine Learning Research*, 158:10235–10264, 2021.

[301] P. Tschandl, C. Rosendahl, and H. Kittler. The ham10000 dataset, a large collection of multi-source dermatoscopic images of common pigmented skin lesions. *Scientific Data*, 5:180161, 2018.

[302] A. Tulsani, P. Kumar, and S. Pathan. Automated segmentation of optic disc and optic cup for glaucoma assessment using improved unet++ architecture. *Biocybernetics and Biomedical Engineering*, 41(2):819–832, 2021.

[303] Z. Ullah, M. Usman, S. Latif, A. Khan, and J. Gwak. Ssmd-unet: semi-supervised multi-task decoders network for diabetic retinopathy segmentation. *Scientific Reports*, 13:9087, 2023.

[304] V7 Labs. Intersection over union (iou): Definition, calculation, code. *V7 Labs Blog*, 2023.

[305] V7 Labs. Mean average precision (map) explained: Everything you need to know. *V7 Labs Blog*, 2023.

[306] D. Valenkova, A. Lyanova, A. Sinitca, R. Sarkar, and D. Kaplun. A fuzzy rank-based ensemble of CNN models for MRI segmentation. *Biomedical Signal Processing and Control*, 102:107342, 2025.

[307] R. Van Berlo et al. Self-supervised learning in medical imaging: a survey of applications and methods. *arXiv preprint arXiv:2309.02555*, 2023.

[308] B. H. M. van der Velden, H. J. Kuijf, K. G. A. Gilhuijs, and M. A. Viergever. Explainable artificial intelligence (xai) in deep learning-based medical image analysis. *Medical Image Analysis*, 79:102470, 2022.

[309] M. Van Rijthoven, M. Balkenhol, K. Siliņa, J. Van Der Laak, and F. Ciompi. Hooknet: Multi-resolution convolutional neural networks for semantic segmentation in histopathology whole-slide images. *Medical Image Analysis*, 68:101890, 2021.

[310] B. VanBerlo and A. Wong. A survey of the impact of self-supervised pretraining for diagnostic tasks in medical x-ray, ct, mri, and ultrasound. *BMC Medical Imaging*, 24(1):79, 2024.

[311] A. Verma, S. N. Shivhare, S. P. Singh, N. Kumar, and A. Nayyar. Comprehensive review on mri-based brain tumor segmentation: A comparative study from 2017 onwards. *Archives of Computational Methods in Engineering*, 31(8), 2024.

[312] M. Voets, K. Møllersen, and L.Å. Bongo. Reproduction study using public data of: Development and validation of a deep learning algorithm for detection of diabetic retinopathy in retinal fundus photographs. *PLoS ONE*, 14(6):e0217541, 2019.

[313] D. Wang, C. Han, Z. Zhang, T. Zhai, H. Lin, B. Yang, Y. Cui, Y. Lin, Z. Zhao, L. Zhao, et al. Feddus: Lung tumor segmentation on ct images through federated semi-supervised with dynamic update strategy. *Computer Methods and Programs in Biomedicine*, 108141, 2024.

[314] J. Wang, H. Ding, F. A. Bidgoli, B. Zhou, C. Iribarren, S. Molloi, and P. Baldi. Detecting cardiovascular disease from mammograms with deep learning. *IEEE Transactions on Medical Imaging*, 36:1172–1181, 2017.

[315] J. Wang, Z. Luan, Z. Yu, J. Gao, J. Ren, K. Khan, K. Yuan, and H. Xu. An adaptive sparse bayesian model combined with joint information-based label fusion for brain tumor segmentation in mri. In *Signal, Image and Video Processing*, pages 1–9, 2021.

[316] W. Wang, E. Xie, X. Li, D. P. Fan, K. Song, D. Liang, T. Lu, P. Luo, and L. Shao. Pyramid vision transformer: A versatile backbone for dense prediction without convolutions. In *Proceedings of the IEEE/CVF International Conference on Computer Vision*, pages 548–558, 2021.

[317] X. Wang, Y. Peng, L. Lu, Z. Lu, M. Bagheri, and R. M. Summers. Chestx-ray8: Hospital-scale chest x-ray database and benchmarks on weakly-supervised classification and localization of common thorax diseases. In *Proceedings of the IEEE conference on computer vision and pattern recognition*, pages 2097–2106, 2017.

[318] Y. Wang, T. Zhou, Y. Lei, S. Bai, and Z. Ge. Transbts: Multimodal brain tumor segmentation using transformer. In *International Conference on Medical Image Computing and Computer-Assisted Intervention (MICCAI)*, pages 109–119. Springer, 2021.

[319] Y. L. Wang, Z. J. Zhao, S. Y. Hu, and F. L. Chang. CLCU-Net: Cross-level connected U-shaped network with selective feature aggregation attention module for brain tumor segmentation. *Computer Methods and Programs in Biomedicine*, 207:106154, 2021.

[320] Z. Wang and J. Yang. Diabetic retinopathy detection via deep convolutional networks for discriminative localization and visual explanation. In *AAAI Workshops*, pages 514–521, 2018.

[321] K. Weiss, T. M. Khoshgoftaar, and D. Wang. A survey of transfer learning. *Journal of Big Data*, 3(9):1–40, 2016.

[322] Wiley Online Library. Image segmentation evaluation with the dice index. *Medical Imaging*, 44(2):123–130, 2024.

[323] T. Y. Wong and C. Sabanayagam. Strategies to tackle the global burden of diabetic retinopathy: from epidemiology to artificial intelligence. *Ophthalmologica*, 243(1):1–12, 2020.

[324] World Health Organization. Cancer facts. https://www.who.int/news-room/fact-sheets/detail/cancer.

[325] X. Wu, L. Bi, M. Fulham, D. D. Feng, L. Zhou, and J. Kim. Unsupervised brain tumor segmentation using a symmetric-driven adversarial network. *Neurocomputing*, 455:242–254, 2021.

[326] C. Xu, X. Guo, G. Yang, Y. Cui, L. Su, H. Dong, X. Hu, and S. Che. Prior-guided attention fusion transformer for multi-lesion segmentation of diabetic retinopathy. *Scientific Reports*, 14:20892, 2024.

[327] L. Xu and Q. Zhao. Hybrid approaches in acl injury diagnostics: Integrating semi-automated and automated methods. *Journal of Medical Imaging and Health Informatics*, 13:234–242, 2023.

[328] C. Xue, L. Zhu, H. Fu, X. Hu, X. Li, H. Zhang, and P. Heng. Global guidance network for breast lesion segmentation in ultrasound images. *Medical Image Analysis*, 70:101989, 2021.

[329] Y. Xue, T. Xu, H. Zhang, L. R. Long, and X. Huang. Segan: Adversarial network with multi-scale l1 loss for medical image segmentation. *Neuroinformatics*, 16(3–4):383–392, 2018.

[330] Y. Yan, P.-H. Conze, G. Quellec, M. Lamard, B. Cochener, and G. Coatrieux. Two-stage multi-scale breast mass segmentation for full mammogram analysis without user intervention. *Biocybernetics and Biomedical Engineering*, 41(2):746–757, 2021.

[331] A. Y. Yang, L. Cheng, M. Shimaponda-Nawa, and H.-Y. Zhu. Long-bone fracture detection using artificial neural networks based on line features of x-ray images. In *2019 IEEE symposium series on computational intelligence (SSCI)*, pages 2595–2602. IEEE, 2019.

[332] F. Ye, Y. Zheng, H. Ye, X. Han, Y. Li, J. Wang, and J. Pu. Parallel pathway dense neural network with weighted fusion structure for brain tumor segmentation. *Neurocomputing*, 425:1–11, 2021.

[333] P. Yin, Y. Xu, J. Zhu, J. Liu, H. Huang, Q. Wu, et al. Deep level set learning for optic disc and cup segmentation. *Neurocomputing*, 464:330–341, 2021.

[334] A. Younis, L. Qiang, C. O. Nyatega, M. J. Adamu, and H. B. Kawuwa. Brain tumor analysis using deep learning and vgg-16 ensembling learning approaches. *Applied Sciences*, 12(14):7282, 2022.

[335] J. S. Yu, C. A. Petersilge, D. J. Sartoris, and M. N. Pathria. Mri of the knee: a comprehensive review. *Radiologic Clinics of North America*, 35(4):776–786, 1997.

[336] X. Yuan, L. Zhou, S. Yu, M. Li, X. Wang, and X. Zheng. A multi-scale convolutional neural network with context for joint segmentation of optic disc and cup. *Artificial Intelligence in Medicine*, 113:102035, 2021.

[337] R. A. Zeineldin, M. E. Karar, Z. Elshaer, J. Coburger, C. R. Wirtz, O. Burgert, and F. Mathis-Ullrich. Explainable hybrid vision transformers and convolutional network for multimodal glioma segmentation in brain mri. *Scientific Reports*, 14:3713, 2024.

[338] X. Zeng, N. Abdullah, and P. Sumari. Self-supervised learning framework application for medical image analysis: a review and summary. *BioMedical Engineering Online*, 23, 2024.

[339] X. Zeng, S. Chen, Y. Xie, and T. Liao. 3V3D: Three-view contextual cross-slice difference three-dimensional medical image segmentation adversarial network. *ACM Transactions on Multimedia Computing Communications and Applications*, 19(6):192, 2023.

[340] X. Zeng, P. Zeng, C. Tang, P. Wang, B. Yan, and Y. Wang. Dbtrans: a dual-branch vision transformer for multi-modal brain tumor segmentation. In *International Conference on Medical Image Computing and Computer-Assisted Intervention*, pages 502–512. Springer, 2023.

[341] J. Zhang, Z. Jiang, J. Dong, Y. Hou, and B. Liu. Attention gate ResU-Net for automatic MRI brain tumor segmentation. *IEEE Access*, 8:58533–58545, 2020.

[342] J. Zhang, J. Zeng, P. Qin, and L. Zhao. Brain tumor segmentation of multi-modality mr images via triple intersecting u-nets. *Neurocomputing*, 421:195–209, 2021.

[343] N. Zhang, Y.-X. Cai, Y.-Y. Wang, Y.-T. Tian, X.-L. Wang, and B. Badami. Skin cancer diagnosis based on optimized convolutional neural network. *Artificial Intelligence in Medicine*, 102:101756, 2020.

[344] P. Zhang. Melanet: A deep dense attention network for melanoma detection in dermoscopy images. In *ISIC*, 2019.

[345] X. Zhang, Y. Wang, C.-T. Cheng, L. Lu, A. P. Harrison, J. Xiao, C.-H. Liao, and S. Miao. Window loss for bone fracture detection and localization in x-ray images with point-based annotation. In *Proceedings of the AAAI Conference on Artificial Intelligence*, volume 35, pages 724–732, 2021.

[346] Y. Zhang, X. Cai, Y. Zhang, H. Kang, X. Ji, and X. Yuan. Tau: Transferable attention u-net for optic disc and cup segmentation. *Knowledge-Based Systems*, 213:106668, 2021.

[347] Y. Zhang, Y. Lu, W. Chen, Y. Chang, H. Gu, and B. Yu. MSMANet: A multi-scale mesh aggregation network for brain tumor segmentation. *Applied Soft Computing*, 110:107733, 2021.

[348] Y. Zhang, B. Zhang, F. Coenen, and L. Wenjin. Breast cancer diagnosis from biopsy images with highly reliable random subspace classifier ensembles. *Machine Vision and Applications*, 24(7):1405–1420, 2013.

[349] F. Zhao, H. Pan, N. Li, et al. High-order brain functional network for electroencephalography-based diagnosis of major depressive disorder. *Frontiers in Neuroscience*, 16:976229, 2022.

[350] C. Zhou, C. Ding, X. Wang, Z. Lu, and D. Tao. One-pass multi-task networks with cross-task guided attention for brain tumor segmentation. *IEEE Transactions on Image Processing*, 29:4516–4529, 2020.

[351] H. Y. Zhou, J. Guo, Y. Zhang, L. Yu, L. Wang, and Y. Yu. nnformer: Interleaved transformer for volumetric segmentation. *arXiv preprint arXiv:2109.03201*, 2021.

[352] Y. Zhou, Z. Chen, H. Shen, X. Zheng, R. Zhao, and X. Duan. A refined equilibrium generative adversarial network for retinal vessel segmentation. *Neurocomputing*, 437:118–130, 2021.

[353] Z. Zhou, Z. He, and Y. Jia. Afpnet: A 3d fully convolutional neural network with atrous-convolution feature pyramid for brain tumor segmentation via mri images. *Neurocomputing*, 402:235–244, 2020.

[354] Z. Zhou, M. M. R. Siddiquee, N. Tajbakhsh, and J. Liang. Unet++: A nested u-net architecture for medical image segmentation. In *Deep Learning in Medical Image Analysis and Multimodal Learning for Clinical Decision Support*, pages 3–11, 2018.

[355] Z. Zhou, M. M. R. Siddiquee, N. Tajbakhsh, and J. Liang. Unet++: Redesigning skip connections to exploit multiscale features in image segmentation. *IEEE Transactions on Medical Imaging*, 39(6):1856–1867, 2020.

[356] J.-Y. Zhu, T. Park, P. Isola, and A. A. Efros. Unpaired image-to-image translation using cycle-consistent adversarial networks. In *Proceedings of the IEEE International Conference on Computer Vision*, pages 2223–2232, 2017.

[357] H. Zou, X. Gong, J. Luo, and T. Li. A robust breast ultrasound segmentation method under noisy annotations. *Computer Methods and Programs in Biomedicine*, 209:106327, 2021.

[358] Ö. Çiçek, A. Abdulkadir, S. S. Lienkamp, T. Brox, and O. Ronneberger. 3d u-net: Learning dense volumetric segmentation from sparse annotation. In *International Conference on Medical Image Computing and Computer-Assisted Intervention (MICCAI)* jau , pages 424–432. Springer, 2016.

[359] I. Štajduhar, M. Mamula, D. Miletić, and G. Uenal. Semi-automated detection of anterior cruciate ligament injury from mri. *Computer Methods and Programs in Biomedicine*, 140:151–164, 2017.

[360] I. Štajduhar, M. Milosavljević, K. Dumičić, J. Kern, and E. Lalić. Semi-automated detection of anterior cruciate ligament injury from mri. *Computer Methods and Programs in Biomedicine*, 155:107–114, 2018.

Index

https://doi.org/10.1515/9783111389059-013

About the authors

Shiv Naresh Shivhare

Dr. Shiv Naresh completed his Ph.D. in Computer Science and Engineering at the Department of CSE, NIT Uttarakhand, and M.Tech at the Department of CSE, MANIT Bhopal, India. He is an Associate Professor in the School of Computer Science at the University of Petroleum and Energy Studies (UPES) Dehradun, India. He is Senior Member, IEEE. His research interests include computer vision and image processing, biomedical image analysis, machine intelligence, and classification.
Email: shiv827@gmail.com

Thipendra P. Singh

Dr. Thipendra P Singh is currently working as Senior Professor of Artificial Intelligence at Chandigarh University, Uttar Pradesh, India. Dr. Singh is also empaneled as an adjunct professor at the School of Computing at Maryam Abacha American University of Nigeria. He holds the prestigious fellowship of the IEI. In addition, he is also a senior member of the IEEE and a member of various other professional bodies, including the ACM, EAI, ISTE, IAENG, etc. He has made more than 120 contributions to various national and international journals. In addition, he has been the author/editor of more than 20 books on various allied topics in computer science. His research interests include machine intelligence, pattern recognition, and the development of hybrid intelligent systems.
Email: thipendra@gmail.com

https://doi.org/10.1515/9783111389059-014

Deepa Joshi

Dr. Deepa Joshi is a Marie Skłodowska-Curie Postdoctoral Fellow under the Horizon Europe programme at the University of Malta. She earned her Ph.D. in Computer Science and Engineering from UPES, Dehradun, India, with research focused on deep learning for medical image analysis. With over nine years of academic and research experience, she works on Multimodal Artificial Intelligence, Machine Learning, Deep Learning, NLP, and Generative AI, particularly for healthcare applications. Her research explores integrating vision, language, and clinical data to build trustworthy and explainable AI systems. Dr. Joshi has authored high-impact journal papers and book chapters and is committed to mentoring students, fostering industry–academia collaboration, and leveraging AI to address realworld challenges in healthcare.
Email: deepajoshi117@gmail.com

Anitesh Mishra

Anitesh Mishra completed his M.Tech in Computer Science and Engineering from Maulana Azad National Institute of Technology (MANIT) Bhopal and B.Tech in Computer Science and Engineering from Institute of Technology & Management, Gorakhpur (UPTU). Currently he is working as an Associate DevOps Architect in R&D Unit Ericsson Charging at Ericsson India Private Limited, Gurgaon, India. He was earlier associated with FIS Global, Amdocs Development Centre, and ATOS India Private Limited and has experience of more than 10 years in Devops technologies and artificial intelligence, machine learning integration, and intelligent automation systems. He is a Microsoft Certified Azure Administrator Associate and Certified Kubernetes Administrator (CKA). His research interests include artificial intelligence, machine learning for predictive diagnostics, intelligent automation, and AI-driven decision support systems.
Email: anitesh.mishra@yahoo.co.in